MATHEMATICAL AND LOGICAL GAMES

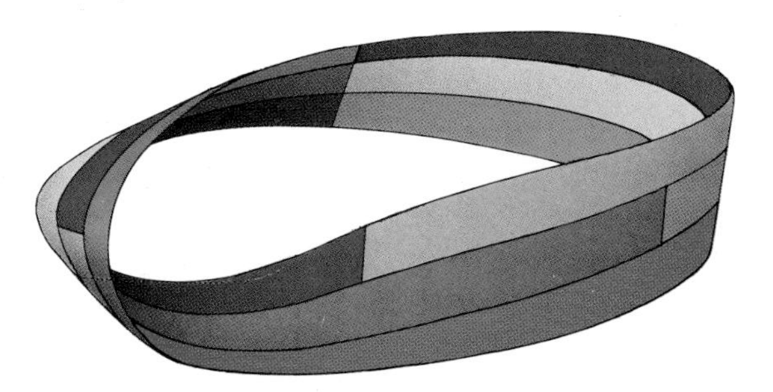

Franco Agostini

MATHEMATICAL AND LOGICAL GAMES

MACDONALD & CO
LONDON & SYDNEY

Colour photographs
Stelvio Andreis and Elvio Lonardi,
Deltaprint S.r.l., Verona
(with the exception of the photograph on p. 35 which is from the Mondadori Archives and the photograph on p. 36 which is by Mario de Biasi)

MATHEMATICAL AND LOGICAL GAMES

A MACDONALD BOOK

Translated from the Italian by Paul Foulkes

First published in Great Britain in 1983 by
Macdonald & Co. (Publishers) Ltd.
London & Sydney

A member of BPCC plc

ISBN 0 356 10086 3

Macdonald & Co. (Publishers) Ltd.,
Maxwell House,
74, Worship Street,
London EC2A 2EN

Printed and bound in Italy by Arnoldo Mondadori Editore, Verona

CONTENTS

INTRODUCTION

"Anyone who regards games simply as games and takes work too seriously has grasped little of either." So wrote the German poet Heinrich Heine a century ago. In today's world the division of work and play persists. Old prejudice still holds that the playing of games is an activity for children, not useless perhaps, but certainly not the responsible and serious work of adults.

Heine intuitively foresaw what modern psychology has since asserted. Games are not only necessary for the development of a child's self-awareness; they are also needed by adults, especially when their work is repetitive and uncreative. The word "games," as it is used in this book, is general and covers a variety of quite complex activities. However, it is those games based on mathematical or logical principles that are among the most absorbing and creative. Indeed, the great mathematicians and scholars of the past often applied their skills to the solution of logical and mathematical games.

This book is a collection of logical and mathematical games both ancient and modern. It is not simply a recital of mathematical pastimes and curiosities set down at random; rational criteria have been used to link the riddles, mathematical problems, puzzles, paradoxes and antinomies. Sometimes the link is an historical reference, at other times it is a conceptual link. Although whenever possible we have repeated some elementary rules, the reader needs no special or complex knowledge. The collection concentrates on some of the simplest and most widely known logical and mathematical games; past events and the people who produced or studied such games are mentioned to make the story more accessible.

Games are presented in a certain sequence. Those mathematical concepts that may be beyond the average reader are stated as clearly as possible. We have omitted complicated games such as chess and draughts and avoided the use of symbolic languages. Technical terms are used only when they are absolutely necessary.

The last two chapters are the most difficult as they include certain abstract philosophical questions involving new mathematical concepts. However, they are rationally linked to what comes before them and they indicate that at critical points in the development of mathematics, problems and logical difficulties were resolved through games that captured the imagination of scholars. Furthermore, even the most abstract mathematical question, logical paradox or antinomy, becomes clearer if it is formulated as a game.

It should be mentioned in passing that games are valuable aids in teaching mathematics; a mathematical device, a riddle or a puzzle, can engage a child's interest more effectively than a practical application, especially when that application is outside the child's experience.

The penultimate chapter offers those readers who may be unversed in logic and mathematics some of the basic concepts and methods on which modern formal logic has been built. The book is designed to inform and educate and is enriched by illustrations of games and riddles that introduce the reader to propositional calculus, the first item of logic with an important place in modern culture. The last chapter, which seeks to complement the one preceding it, begins with ordinary games of chance to demonstrate the importance of the concept of mathematical probability to an understanding of various objective and subjective facts.

Because the last two chapters tend to be abstract and difficult, we have added a list of concrete problems, examples, and games, along with their solutions, to permit the reader to test his grasp of the more theoretical. We happen to believe that the solutions to such games and problems, even if elementary, would be hard to reach without an adequate theoretical foundation. The book concludes with an extensive bibliography to guide the reader in the search for new games and a deeper insight into historical and philosophical questions.

GAMES WITH NUMBERS

. . . The symbolic language of mathematics is . . . a kind of brain-relieving machine on which we easily and often perform symbolic operations that would otherwise tire us out. (Ernst Mach D. Reider)

An historical note

From our earliest schooling we have been taught to operate with whole numbers, fractions, negative numbers and the like; perhaps only a few of us, however, have asked ourselves what numbers are or represent. Numbers developed with man and have marked his life from the beginning of civilization. Indeed, the world is based on numbers. Numbers began as symbols invented by man for a variety of uses, perhaps most immediately for counting the elements in sets of things. "2" could mean two cows or any other two things: two donkeys, two rocks. "2 + 3," assuming "3" meant three donkeys, could represent the sum of two cows and three donkeys, or it could represent a different kind of set—two things of one kind and three of another. Numbers are mental constructions that can indicate material objects without noting their particular features. They are instruments that enable us to make rapid calculations and present quantitative expressions in a simple synthetic way.

In the course of history different peoples have used different symbols to represent numbers. The ancient Romans, for example, indicated "two" as II, "three" as III; while V, the sign for "five," symbolized the five fingers of one hand, and the sign X symbolized two hands, one across the other, for twice five. We have since adopted another set of symbols of Indo-Arabic origin. Why? At first it might even seem that these later symbols are more complicated. Is it not easier to grasp I, II, III, and V, than 1, 2, 3, and 5? Yet the Arabic notation has displaced the Roman one, primarily due to the practices of medieval Italian merchants, and particularly to the influence of the Pisan mathematician Fibonacci, born in 1179. In fact, the mathematician's name was Leonardo da Pisa, however he acquired the nickname Fibonacci because he was the "son of Bonacci," a well-known merchant and official in 12th-century Pisa. The elder Bonacci traded with Arab countries in North Africa and the East, and was accompanied by his son on his frequent trips; hence Fibonacci attended Muslim schools and adopted their algebraic methods together with the Indo-Arabic system of numerals. He later recorded his education in arithmetic, algebra and geometry in his book *Liber Abaci* (1202), and demonstrated the simplicity and practicality of the Indo-Arabic system as opposed to the Roman system.

Greek numerical notation

α′ = 1	τ′ = 300
β′ = 2	υ′ = 400
γ′ = 3	φ′ = 500
δ′ = 4	χ′ = 600
ε′ = 5	ψ′ = 700
ς′ = 6	ω′ = 800
ζ′ = 7	ϡ′ = 900
η′ = 8	,α = 1000
θ′ = 9	,η = 8000
ι′ = 10	,ξ = 60,000
ϰ′ = 20	,ϟ = 90,000
λ′ = 30	,ϱ = 100,000
μ′ = 40	etcetera
ν′ = 50	ιβ′ = 12
ξ′ = 60	μθ′ = 49
ο′ = 70	ϱλα′ = 131
π′ = 80	ωε′ = 805
ϟ′ = 90	,αυϰ′ = 1420
ϱ′ = 100	,ι,γϱα′ = 13,101
σ′ = 200	etcetera

Roman numerical notation

I = 1	M = 1000
II = 2	LX = 60
III = 3	DC = 600
IV = 4	XL = 40
V = 5	XC = 90
VI = 6	CD = 400
VII = 7	CM = 900
VIII = 8	MM = 2000
IX = 9	$\overline{\text{II}}$ = 2000
X = 10	$\overline{\text{C}}$ = 100,000
L = 50	$\overline{\|\text{X}\|}$ = 1,000,000
C = 100	1983 = MCMLXXXIII
D = 500	

Left: The Greek number system, consisting of the 24-letter alphabet and the three signs: stigma, ϛ′, for the number 6, koppa, ϟ′, for 90, and sampi, ϡ′, for 900. The system is decimal, with letters for units, tens and hundreds.

In the West, however, many men of science, trade and letters opposed the "new fashion," and it was a while before it took root. In Florence, for instance, the Statutes of the Art of Exchange prohibited bankers from using Arabic numerals. On the whole, people were hostile to the Arabic system as it made reading commercial records more difficult, but in time the new fashion established itself. The reasons it did are linked to the nature of mathematics itself, namely simplicity and economy. A symbolic system of ten signs (0, 1, 2, 3, 4, 5, 6, 7, 8, 9) serves to represent any number, however large or small, because in the representation of numbers, the meaning of the numerals changes according to their position. Thus, in the number 373, the two numerals 3, though the same symbol, mean different things: the first indicates hundreds, the last indicates units, while the 7 indicates tens. There is no other symbolic system so simple or effective.

The Romans, and before them the Egyptians, Hebrews and Greeks, used a clumsy numerical system based only on a principle of addition. The Roman number XXVIII, for example, means ten + ten + five + one + one + one. The expression of numbers by a few symbols that change in meaning according to position was apparently used by the Chaldaeans and Babylonians of Mesopotamia. Later it was developed by the Hindus who transmitted it to the Arabs and they, in turn, passed it on to mathematicians of medieval Europe.

The introduction of Indo-Arabic numerals with positional notation greatly influenced subsequent developments in mathematics. It simplified mathematical concepts and freed them from the encumbrances stemming from representing mathematical operations in material terms. The Greeks and Romans, for example, used complex geometric systems for multiplication; hence the concept of raising a number to a power (as product of so many equal factors) could not be understood or made simple, especially when dealing with numbers raised to powers higher than the third. To illustrate (Fig. 1), if the number three represents a line three units long, and $3 \times 3 = 3^2$ represents an area, and $3 \times 3 \times 3 = 3^3$ a volume, what meaning might be attached to $3^4 =$

Opposite, right: The Roman number system, also decimal, used fewer symbols: I, V, X, L, C, D, M, for 1, 5, 10, 50, 100, 500, 1,000 respectively. Multiples of units, tens, hundreds, thousands are repeated up to four times (thus 3 = III, 200 = CC). A number to the right of a bigger one is understood as added, thus LX = 60, DCC = 700. A number to the left of a bigger one is understood as subtracted, thus XL = 40. A horizontal line over a number multiplies it by 1,000, for example $\overline{C}$ = 100,000. A number enclosed by $\overline{|\;\;|}$ is multiplied by 100,000; thus $\overline{|X|}$ = 1,000,000. This system spread throughout the Roman world and persisted until it was replaced in the 13th century by the Indo-Arabic system.

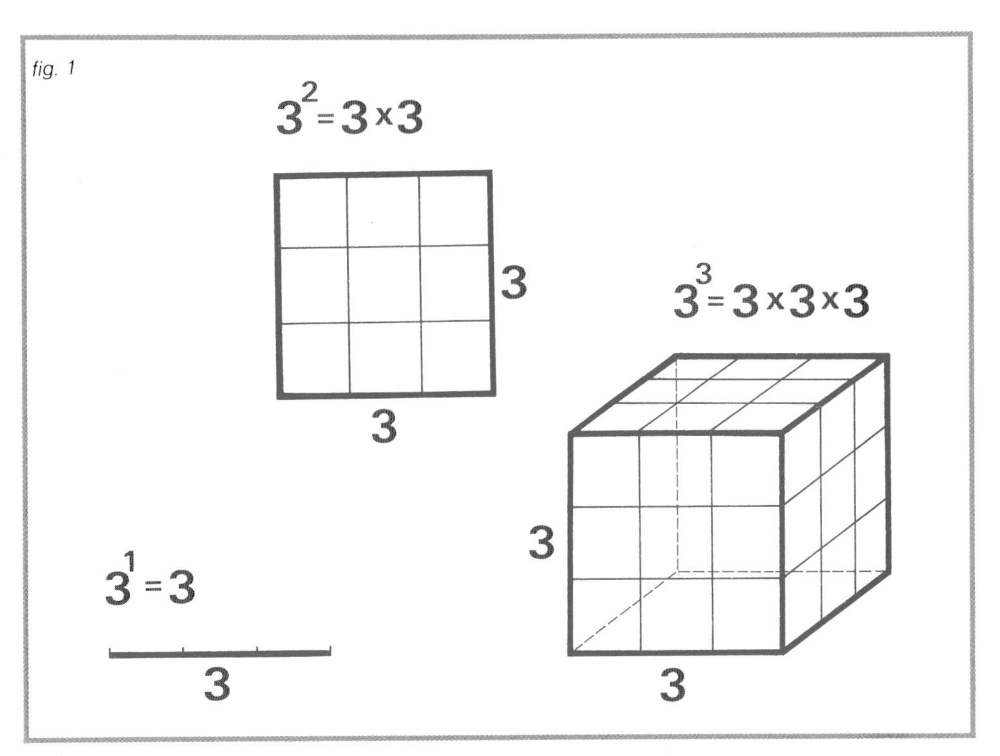

$3 \times 3 \times 3 \times 3$ or $3^5 = 3 \times 3 \times 3 \times 3 \times 3$? Using Indo-Arabic numerals, we find that they are simply numbers.

A first curiosity

To be sure, man first used numbers to solve his practical problems more quickly, but we like to think he also used them to entertain himself. On this assumption, we shall begin our book with a rather popular game that requires only the most elementary numerical calculations. Take the set of digits 1, 2, 3, 4, 5, 6, 7, 8, 9. The game is to insert symbols for mathematical operations between the numerals so the result will equal 100. We are not allowed to change the order of the digits.

Here is one possible solution:

$$1 + 2 + 3 + 4 + 5 + 6 + 7 + (8 \times 9) = 100.$$

In the last part of the expression we have used multiplication, but the game is more interesting if the operations are confined to addition and subtraction. Here is a solution:

$$12 + 3 - 4 + 5 + 67 + 8 + 9 = 100$$

The reverse game can also be played, with the digits decreasing in order: 9, 8, 7, 6, 5, 4, 3, 2, 1. Now reach a sum of 100 using the fewest "+" and "−" signs. A possible solution is:

$$98 - 76 + 54 + 3 + 21 = 100$$

If you are familiar with the properties of numbers, you can solve the following as well. Find three positive integers whose sum equals their product. One solution is:

$$1 \times 2 \times 3 = 1 + 2 + 3 = 6$$

Note that 1, 2, 3 are the factors of 6, which is their sum. We continue the game by finding the number after 6 equal to the sum of its factors. The number is 28, as the factors of 28 are 1, 2, 4, 7, 14 and:

$$1 + 2 + 4 + 7 + 14 = 28$$

Such numbers form a series (after 28 comes 496) called "perfect numbers." It was the mathematician Euclid, famous for his *Elements* of geometry and a resident of Alexandria during his most active years (306–283 B.C.), who first created a

Archimedes' spiral (*below*) and natural spirals (*left*, a nautilus shell in sections) can be expressed by Fibonacci numbers.

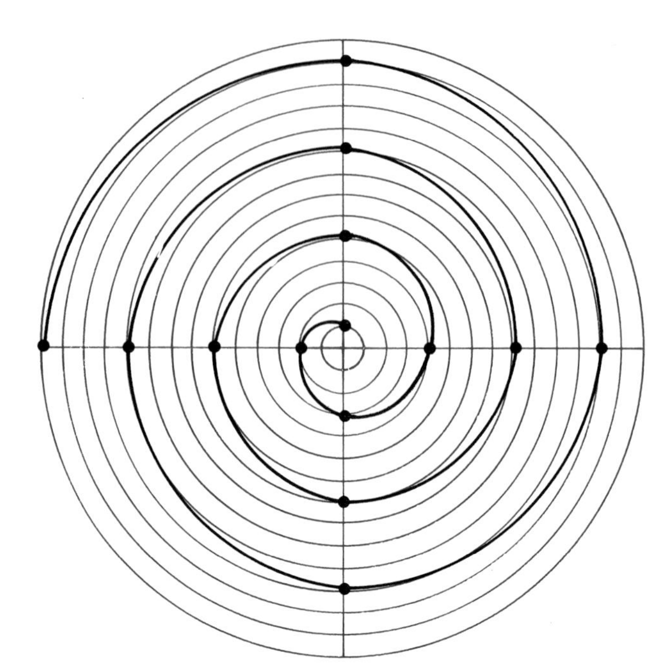

formula for the structure of perfect numbers, namely:

$$N = 2^{n-1}(2^n - 1)$$

In this formula, the second factor, $(2^n - 1)$, must be a prime number, that is, divisible only by itself and unity. Thus n must be such that $2^n - 1$ is prime. It is easy to see that the latter is not prime if n is not prime. The reader should try to use this formula to find the next perfect number after 496. After that, the calculations become rather lengthy.

Here is a table for the first nine perfect numbers:

	n	2^{n-1}	$2^n - 1$	Perfect numbers
1	2	2	3	6
2	3	4	7	28
3	5	16	31	496
4	7	64	127	8128
5	13	4096	8191	33550336
6	17	65536	131071	8589869056
7	19	262144	524287	137438691328
8	31	1073741824	2147483647	2305843008139952128
9	61	–	–	2658455991569831744654692615953842176

We observe that all perfect numbers obtained by Euclid's formula are even and always end in 6 or 8.

Fibonacci numbers

Among the many arithmetical and algebraic questions studied by Fibonacci, that of *sequences* deserves special attention, as it was the basis for his interesting problem of *the rabbits*. Suppose we put a pair of adult breeding rabbits in a cage to produce offspring, and that each month they produce another pair, which, in turn, breed after two months. (This is hypothetical, of course, as rabbits do not reach maturity before four months.) If all the rabbits survive, how many will there be at the end of one year? The solution is indicated in Fig. 2. We start in January with the initial pair A. In February there will be two pairs, A and their offspring B. In March A produces C, which makes three. However, in April, A produces D, while B, now mature, produces E. In May, it is more complicated still: A produces F, B produces G, and C produces H. Continuing in this fashion,

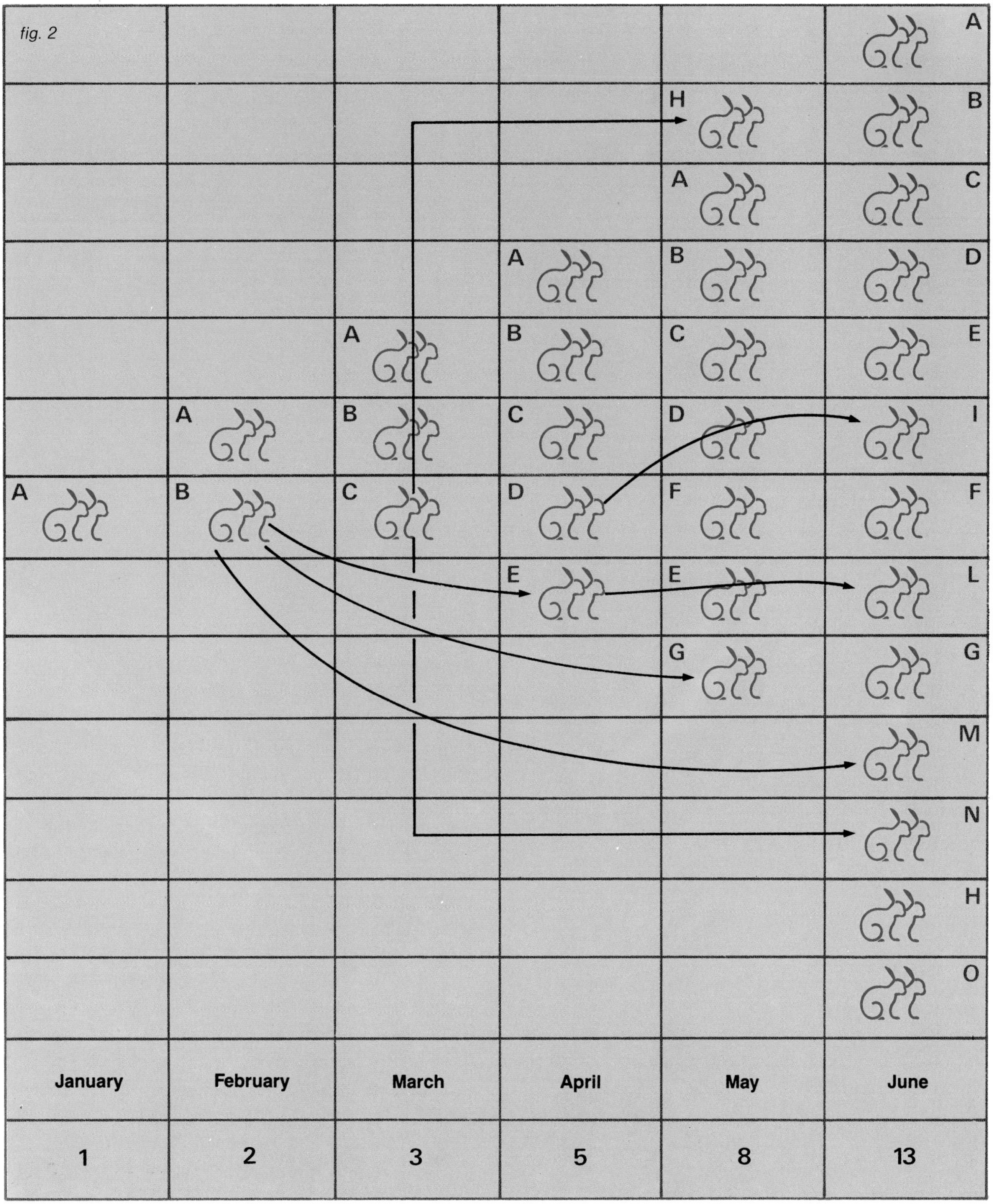

fig. 2

					A
				H	B
				A	C
			A	B	D
		A	B	C	E
	A	B	C	D	I
A	B	C	D	F	F
			E	E	L
				G	G
					M
					N
					H
					O
January	**February**	**March**	**April**	**May**	**June**
1	2	3	5	8	13

the number of pairs produced in successive months is: 1, 2, 3, 5, 8, 13. . . .

The law linking the numbers is easily detected. From 3 on, each number is the sum of the two

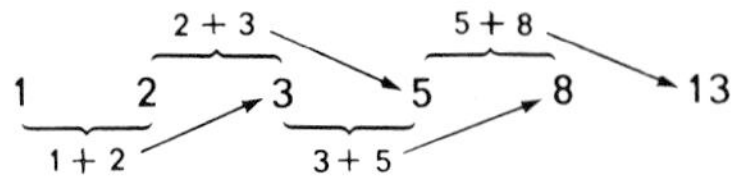

2 + 3 5 + 8
1 2 3 5 8 13
1 + 2 3 + 5

preceding it. Hence we can easily find the numbers for the later months: July, 8 + 13 = 21; August, 13 + 21 = 34; September, 21 + 34 = 55; October, 34 + 55 = 89; November, 55 + 89 = 144; December, 89 + 144 = 233. At the end of the year there will be 233 pairs. Once the formation law is found, the sequence can be continued indefinitely.

Fibonacci did not explore the question of number sequences more deeply, and it was not until the 19th century that mathematicians began to study their formal properties. In particular, François Edouard Anatole Lucas investigated the *Fibonacci series*, where starting with any two integers the next term is the sum of the two before. The table shows the first twenty terms of the series starting 1, 1 and 1, 3.

1	1
1	3
2	4
3	7
5	11
8	18
13	29
21	47
34	76
55	123
89	199
144	322
233	521
377	843
610	1 364
987	2 207
1 597	3 571
2 584	5 778
4 181	9 349
6 765	15 127

Fibonacci series have always captured the imagination of mathematicians and enthusiasts who have tried endlessly to unearth their hidden properties and theorems. Recently, such series have been useful in modern methods of electronic programming, particularly in data selection, the recovery of information, and the generation of random numbers.

A curious calculating device: the abacus

Man has always tried to do sums with greater speed. The Babylonians cut permanent signs on clay tablets to hasten calculations. Subsequently, the abacus was invented—where and when is not known, perhaps in ancient Egypt. The abacus was the first calculating machine and it was an ingenious instrument. Numbers were represented as objects (pebbles, fruit stones, and pierced shells for example) and placed on small sticks fixed to a support. The word *abax, abakos* means a "dust-covered tablet" on which geometrical figures can be traced or calculations performed, and it probably came into ancient Greek from the Hebrew *abaq,* meaning "dust." Thus the word originated in the Near East.

Although the mathematicians of ancient Greece were familiar with the discoveries of Mediterranean peoples, and enhanced them with original notions of their own, their mathematical advances had no discernible impact on the structure or workings of Greek society. Indeed, such advances were seen as little more than intellectual exercises. We know, too, that new scientific and technical knowledge were seldom used to achieve greater productivity or freedom from physical labour; again they were treated simply as expressions of man's creative ability. This prejudice impeded the progress of mathematics in Greece and explains why many of the major arithmetical and algebraic discoveries came to us

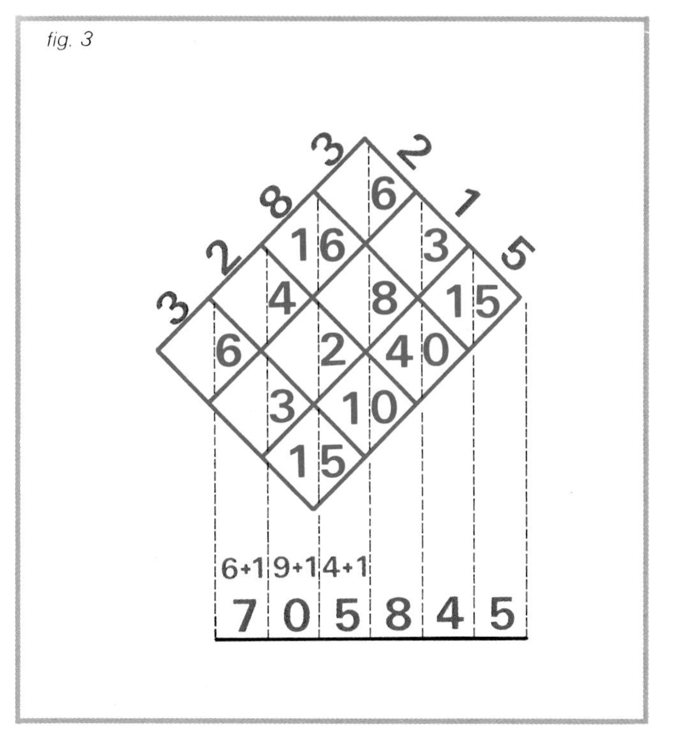

fig. 3

fig. 4

from the Indian and Arab algebraists working between A.D. 400 and 1200. Their discoveries were brought to Italy and thence to the West by the traders in the maritime republics.

Mathematical knowledge spread widely after the Protestant Reformation and the invention of paper and the printing press. Indeed, it was Martin Luther who insisted that the first arithmetic textbooks be printed. Indian algebraists, and later the Arabs, had demonstrated the advantages of the new positional number system; calculations could be simplified much to the relief of those who used them in trade and commerce. Again, it was through commerce that the abacus found its way to the West, and today this simple calculator is still used in Russia, China and Japan to total bills in shops and restaurants, and frequently to teach arithmetic.

Abaci have varied between peoples and periods. The abacus with beads on small sticks (pp. 18–19) is only one type, probably of Chinese origin. The Arabs developed others of a different construction and one still in use consists of a kind of grid. It is best explained by carrying out a multiplication, say 3,283 by 215. Draw a rectangle of as many small squares as the two factors have figures, in this case 4×3, with their diagonals vertical and horizontal. Divide the unit squares vertically and extend the traces to a base line (Fig. 3). We have put the two factors on the sides—the four-figure number on the longer side and the three-figure number on the shorter side. The result is a grid which is now filled with the products of the figures at the edge. For example, the one farthest right contains $5 \times 3 = 15$, the one farthest left $2 \times 3 = 6$; the units are in the right division and the tens in the left division. When the grid has been filled, we add on the base line, carrying tens as needed. The result is 705,845, which can be checked by the ordinary method.

Fig. 4 shows a similar abacus on which 1,176

(continued on page 20)

What numbers are.
The photographs (*left*) show three Munich beer mats, on which waiters mark each beer ordered by the customer. The bill is then tallied by simply multiplying the price of one glass times the number of marks on the mat. Such methods are ancient and date back to the mythical origins of mathematics.

The notion of a mark corresponding to a unit eventually produced the natural, or counting, numbers. In order to simplify counting operations a special set of symbols was gradually devised, namely what are known as ciphers, figures or numerals. Natural numbers were used in barter (exchanging one article for another) and in childrens' games where objects were classified, put in sequence or set out according to rules of proximity, continuity and boundaries.

From the natural numbers and the operation of adding, man has gradually constructed the entire system of numbers (cf pp. 21, 104) as well as the other calculating operations (note that, formally, the positive integers are ratios of natural numbers to the natural number 1, a definition that was not put forth until the 19th century).

Opposite: A register used by illiterate Sicilian shepherds. By notching sticks the shepherds indicated the number of animals—sheep or goats—each owned, the animals' births, the dairy products produced and so on. This calendar, valid from September 1st for one year, was settled every August 31st. As the date approached, a literate man was appointed to inscribe each stick with certain signs indicating its owner's name and the type of animals he possessed:

heads of beasts each member owned,

births

males

dairy products, curd, cheese, etc.

While the number of adult animals was almost constant for a year, dairy products and new births varied from month to month (males were sold, females were kept), so the register was updated every month.

Although the people who used this instrument could neither read nor write, they could count intuitively and they correlated signs and objects in a highly complex fashion. Instruments such as this form the basis of mathematical and logical thinking.

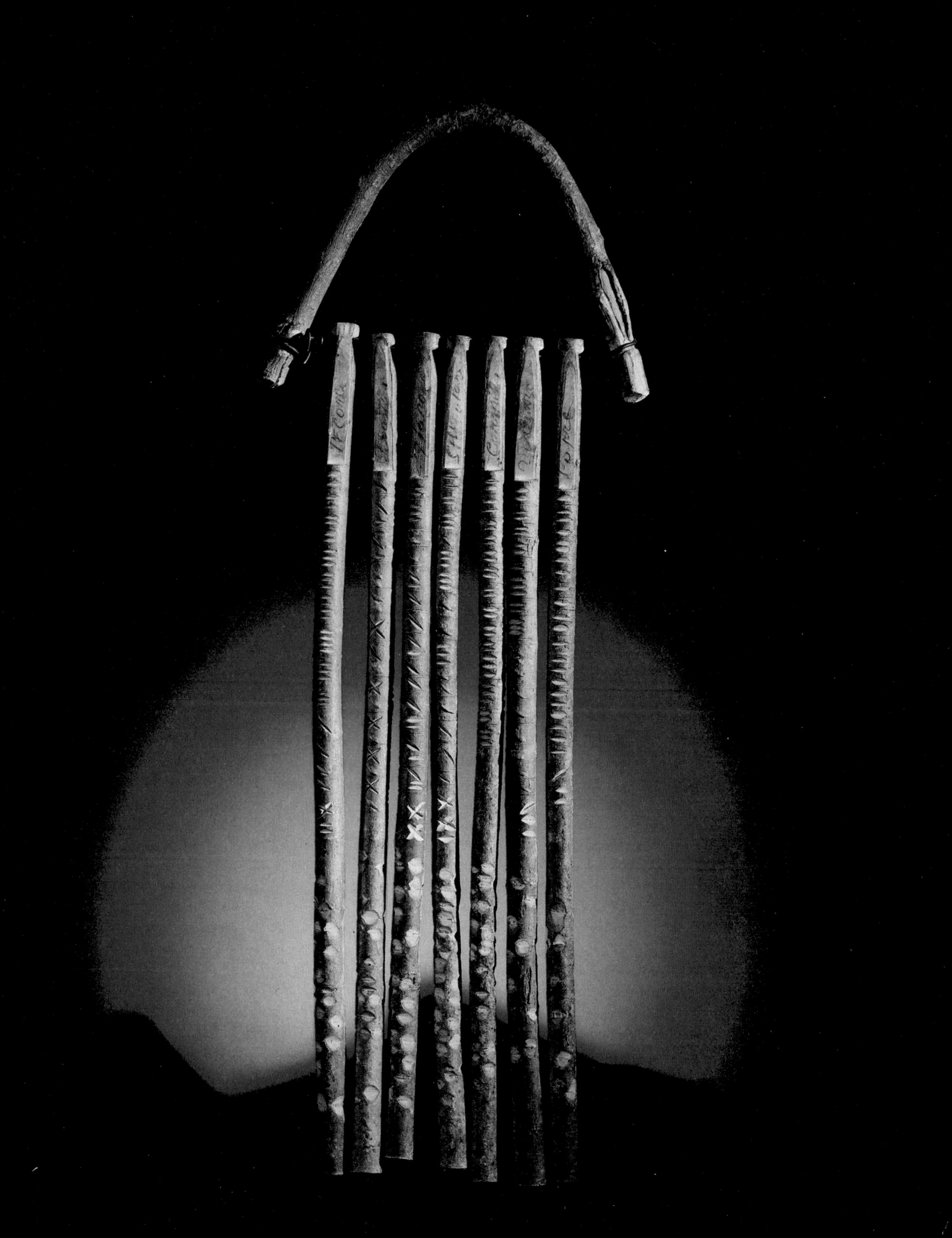

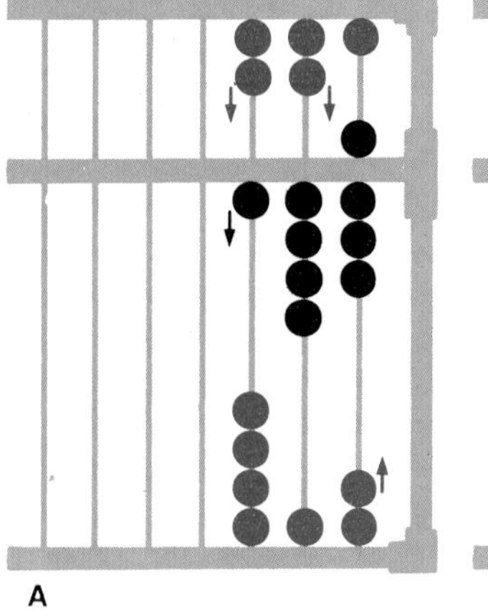

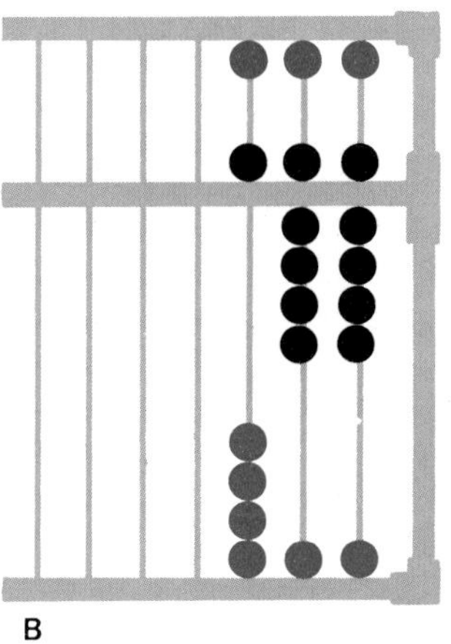

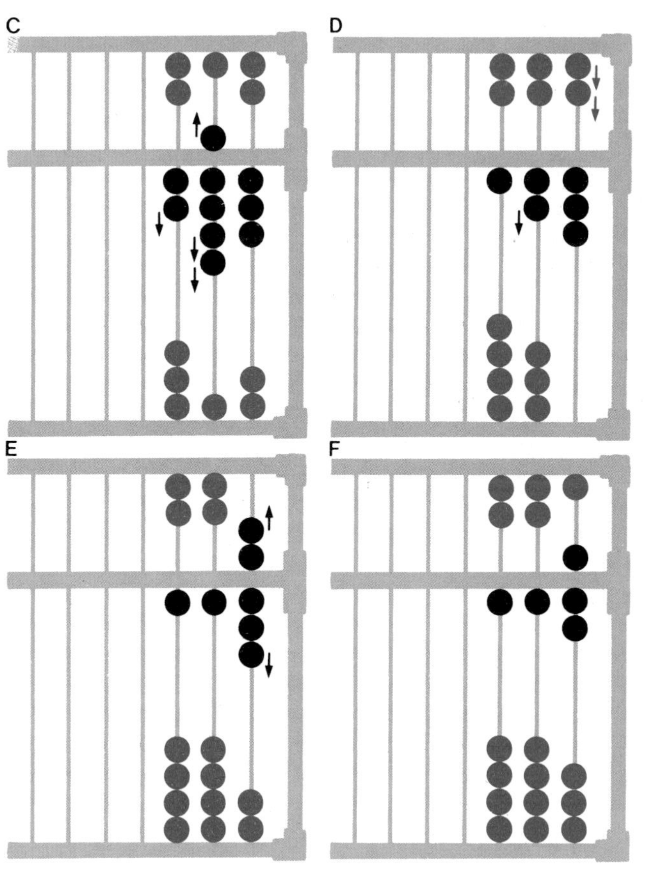

Addition and subtraction on the abacus.
This abacus (*opposite*) is currently used in China (the whole instrument appears on the book jacket). Each vertical column of beads, starting on the right with the units, indicates the digits of numbers in positional notation. Each bead under the bar stands for 1, 10, 100 and so on. Each bead on the bar stands for 5 of the items in its column. Thus the number shown here is 173.

To add, start from the right. Suppose we need to add 148 and 451. Form the number 148 (A), add one unit, five tens and four hundreds producing 599 (B). To subtract, we start on the left and do the reverse. Take 293 – 176. Form 293 (C), take away one hundred, and seven tens—by removing the 50 and two 10s—(D); to subtract six units, borrow a ten in the unit column by taking away one ten in the column of tens and adding two fives in the unit column (E). Now we can take away six units, and the result is 117 (F).

and 6,895 can easily be multiplied. First, write the product of 1 by 6,895 in the first row (tens in the top half and units in the bottom half). Put the product $1 \times 6{,}895$ in the second row, $7 \times 6{,}895$ in the third row, and $6 \times 6{,}895$ in the fourth row. Sum diagonally from the right and carry to the left as needed. A schema much like this served Pascal and Leibnitz four centuries later when they conceived the idea of the first calculating machines.

The origins of algebra

The term "algebra" derives from the Arabic *al-jebr,* which the mathematician Al-Khowarizmi adopted to explain his ideas for solving what we call equations. Later the term acquired a wider meaning and today it includes a broad range of mathematics.

Mohammed ibn Musa Al-Khowarizmi, an Arabian astronomer and mathematician (died ca.A.D. 850), was active in the "House of Wisdom" in Baghdad, a cultural center established about A.D. 825 by the Caliph Al-Mamun. Al-Khowarizmi wrote various books on arithmetic, geometry and astronomy and was later celebrated in the West. His arithmetic used the Indian system of notation. Although his original Arabic book on the system, probably based on an Indian text, is lost, a Latin translation survives as *Algorithmi: De numero indorum* (about Indian numbers). The author explains the Indian numerical system so clearly that when the system eventually spread through Europe, it was assumed the Arabs were its inventors. The Latin title gives us the modern term "algorithm"—a distortion of the name Al-Khowarizmi which became Algorithmi—used today to denote any rule of procedure or operation in calculations.

Al-Khowarizmi's most important book, *Al-jebr wa'l-muqabalah,* literally "science of reducing and comparing," gave us the word "algebra." There are two versions of the text, one Arabic and the other the Latin *Liber algebrae et almucabala* which contains a treatment of linear and quadratic equations.

These works were of major importance in the history of mathematics. Indeed, *al-jebr* originally meant a few mathematical steps and transformations to simplify and hasten the resolution of problems.

Let us now turn to what we learned in school and begin with an equation of the first degree, $5x + 1 = 3(2x - 1)$. An equation is generally an equality with one or several unknowns. It translates into numbers a problem whose solution consists of finding those values of x that make the equality true. In our example, we must find the value of x that makes the expressions on either side of the "equal" sign equal.

Al-Khowarizmi's mathematical works contain all the solving procedures we learned mechanically in school, for example, *reducing* terms and *transferring* a term to the other side with a change of sign. Hence, in our case, adding 3 and subtracting $5x$ on both sides, and then changing sides, gives us $x = 4$, which solves the equation. Putting 4 for x in the first equation, $5x + 1 = 3(2x - 1)$, we find $21 = 21$. Clearly, to solve an equation is to transform it into other, and simpler equations until we reach the solution.

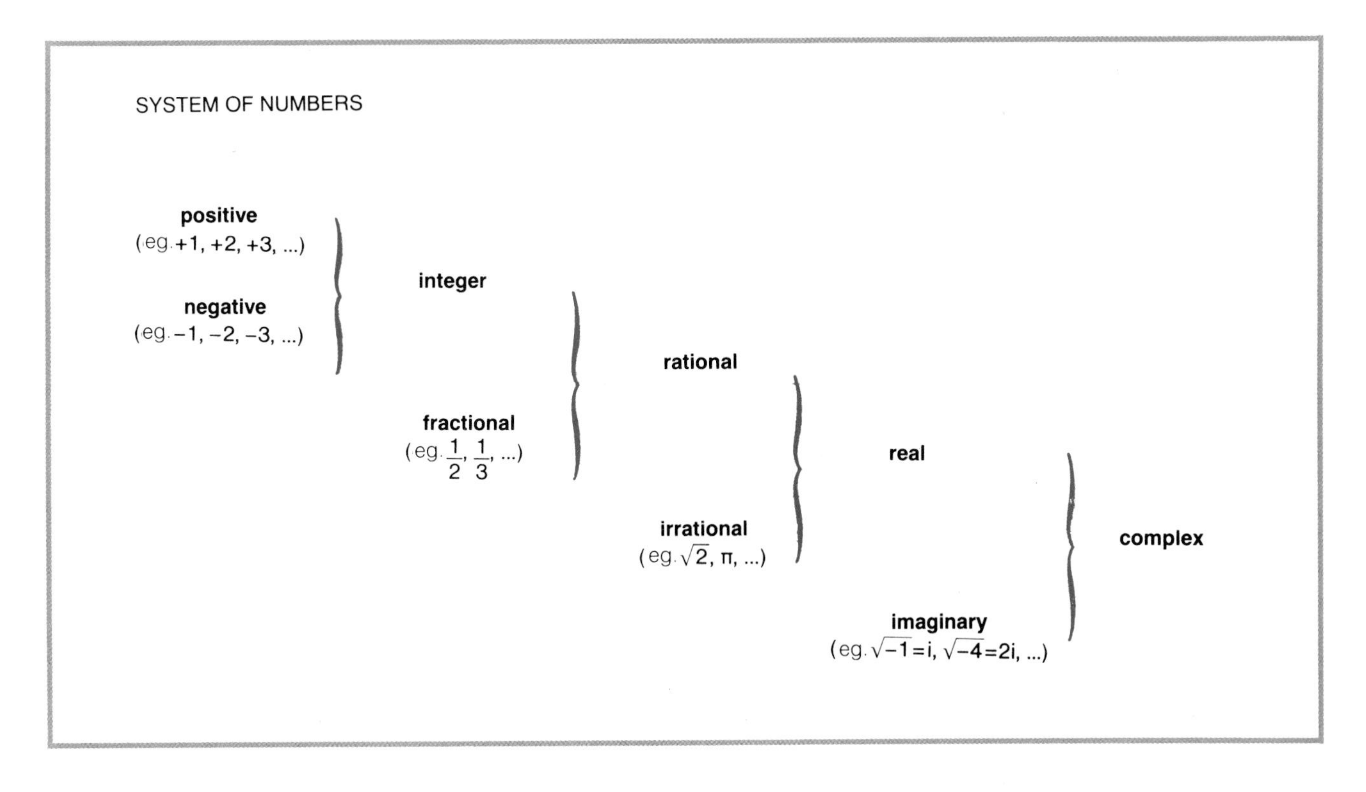

Games with algebra

Algebra and its laws have often spawned tricks and games that seem to smack of magic. In fact, they are readily explained by algebraic laws. Imagine that we have asked someone to play this game:

1) Think of a number;

2) add 3;

3) multiply by 2;

4) subtract 4;

5) divide by 2;

6) subtract the original number.

Whatever the original number, the result must always be 1. Surprising? Not if you give the simple algebraic explanation. The principles are elementary but deep. Take the present game; the result does not depend on the original number, which need not be known. The expression "any number" can mean two different things in algebra, either a *variable* number which can have various values, or a given number which is undetermined, namely a *constant* whose value has not been fixed. For clarity, variables are shown by the last letters of the alphabet ($x, y, z \ldots$), and constants by the first letters ($a, b, c \ldots$). Thus if we write $3+x$, with x integer and variable, we have for $x=-1$, $3+(-1)=2$; for $x=0$ we have $3+0=3$; for $x=1$, $3+1=4$.

A variable in an equality, say x in $6=5+x$, becomes an *unknown* (a value not at first known), indicating the value required to verify the equality. Returning to our game:

1) Take a number, x;

2) add 3, $x+3$;

3) multiply by 2, $2(x+3)$;

4) subtract 4, $2(x+3)-4$;

5) divide by 2, $(2(x+3)-4)/2$;

6) subtract the original number x, $(2(x+3)-4)/2-x$. In algebra, this last expression represents the sequence of verbal moves. Whatever x, this expression equals 1.

An expression such as $(2(x+3)-4)/2-x=1$ is called an *identity*. The difference between this and an equation is readily explained. In an identity the two sides are always equal, whatever the value of x, while in an equation this is not so. Returning to our first equation, $5x+1=3(2x-1)$, there is only one value of x for which the two sides are equal, namely the single solution of that linear equation, which, as we saw, is $x=4$. For any other value of x, the sides are unequal. For example, if we put $x=0$, we have $1 \neq -3$ (the sign means "different from").

By applying such elementary principles of algebra other games can be invented. For example, the following always results in 5:

1) Take a number, x (say, 6);

2) add its successor, $x+(x+1)$ (here, $6+7=13$);

3) add 9, $2x+10$ (here, $13+9=22$);

4) divide by 2, $(2x+10)/2$ (here, $22/2=11$);

5) subtract the original number, $x+5-x=5$ (here, $11-6=5$).

	(A): $2(x+3)-4=2+2x$	is an identity;	(B): $5x+1=3(2x-1)$	is an equation
if we set $x=0$	in (A) we obtain	$2=2$;	in (B) we obtain	$1 \neq -3$
if we set $x=1$	in (A) we obtain	$4=4$;	in (B) we obtain	$6 \neq 3$
if we set $x=2$	in (A) we obtain	$6=6$;	in (B) we obtain	$11 \neq 9$
if we set $x=3$	in (A) we obtain	$8=8$;	in (B) we obtain	$16 \neq 15$
if we set $x=4$	in (A) we obtain	$10=10$;	in (B) we obtain	$21=21$
if we set $x=5$	in (A) we obtain	$12=12$;	in (B) we obtain	$26 \neq 27$

The algebraic expression reduces to 5, whatever x. The game is a bluff, the trick lying in the intricate instructions which are designed to complicate matters. Actually each step is simple, but we are not always alert enough to see the main points. In the present instance, the trick is to take away the arbitrary original number, that is the subtraction $x - x$. By starting the whole process from the other end, any number of such games can be invented. Let us construct one that always results in 13. For any real number there are infinite identities. Take $13 = 7 + 6$, for example. Since $x - x = 0$, we can add this to the right-hand side without upsetting the identity, $13 = 7 + 6 + x - x$. This can now be rewritten as $13 = (2(7 + 6 + x))/2 - x$ because multiplying and then dividing an expression by the same number leaves it unchanged. Next we can make things more complex by multiplying out the bracket: $13 = (14 + 12 + 2x)/2 - x$, which can be recast as $13 = (2(x + 7) + 12)/2 - x$. The game then is this:

1) Take a number, x (suppose we take 10);

2) add 7, $x + 7$ (here, $10 + 7 = 17$);

3) multiply by 2, $2(x + 7)$ (here, $2 \times 17 = 34$);

4) add 12, $2(x + 7) + 12$ (here, $34 + 12 = 46$);

5) divide by 2, $(2(x + 7) + 12)/2$ (here, $46/2 = 23$);

6) take away the original number, $(2(x + 7) + 12)/2 - x$ (here, $23 - 10 = 13$).

The rules of algebra are such that it can appear we are able to read peoples' minds. Try this exercise:

1) Think of a number, x (suppose we take 6);

2) double the number, $2x$ (here, $2 \times 6 = 12$);

3) add 4, $2x + 4$ (here, $12 + 4 = 16$);

4) divide by 2, $(2x + 4)/2$ (here, $16/2 = 8$);

5) add 13, $(2x + 4)/2 + 13$ (here, $8 + 13 = 21$).

We now ask the player for his answer, namely 21, and quickly tell him he started with 6. Since the final algebraic expression reduces to $x + 15$, we know that, in this instance, $x + 15 = 21$, so $x = 6$. Algebra, not mind reading.

There are countless variations of this game, as the natural numbers that can be subtracted from x, which is itself a natural number, are endless. The number subtracted should not be too small, say at least 20. The first step would be $x + 20$, an expression which will now be transformed. For example:

$$x + 20 = x + 6 + 14 = 3(x + 6)/3 + 14 = (3x + 18)/3 + 14.$$

The game then consists of these steps:

1) Think of a number, x (say, 8);

2) multiply by 3, $3x$ (here, $8 \times 3 = 24$);

3) add 18, $3x + 18$ (here, $24 + 18 = 42$);

4) divide by 3, $(3x + 18)/3$ (here, $42/3 = 14$);

5) add 14, $(3x + 18)/3 + 14$ (here, $14 + 14 = 28$);

6) subtract 20, $(3x + 18)/3 + 14 - 20 = x$ (here, $28 - 20 = 8$); which produces the number originally in mind.

Odds and evens

From our early efforts in arithmetic we learned to distinguish between odd and even numbers; the latter are divisible by 2, the former are not.

Let us examine the algebraic notation of an even number and its properties. Take any integer x; $2x$ then is even. Thus 14 is even, for we can write it as 2×7. If $2x$ is even, then $2x + 1$ is odd. For example, $15 = (2 \times 7) + 1$.

This hatches some amusing games. Let a player take an even number of coins in one hand and an odd number in the other. Ask him to double the number of coins in his left hand and triple the number in his right, and to reveal the total of the two numbers. You can then tell him which hand holds the odd number of coins and which hand has the even number. If the sum is odd, the odd number of coins is in the right hand; if the sum is even, the odd number of coins is in the left hand. For example: if we have three coins in the left hand and six in the right then $2 \times 3 + 3 \times 6 = 24$. The sum is even, and the odd number of coins is in the left hand. What is the trick? We need algebra to grasp it. As before, we can follow the operations step-by-step. There are two possibilities:

1) The odd number of coins is in the left hand.

2) The odd number is in the right hand.

Call the number of coins in the left hand L and the number in the right hand R. Then:

1) Odd number of coins in the left hand, $L = 2x + 1$, $R = 2y$, where x, y are two unknown integers whose actual value does not matter. The sum to be considered is $2L + 3R = 4x + 2 + 6y = 2(2x + 1 + 3y)$ which is divisible by 2, and hence even.

2) Even number of coins in the left hand, $L = 2x$, $R = 2y + 1$, and $2L + 3R = 4x + 6y + 3 = 2(2x + 3y + 1) + 1$, which is odd. This completes the proof.

The successor of a number

Those algebraic expressions that almost look too simple at first, can actually suggest a variety of entertaining mathematical games—successive numbers, for example. These are numbers that come directly after one another: $x + 1$ follows x, $x + 2$ follows $x + 1$, and so on.

Take five successive numbers and add them together:

$$x + (x + 1) + (x + 2) + (x + 3) + (x + 4) = 5(x + 2)$$

This produces the next game.

1) Tell someone to think of a number, x (suppose it is 252);

2) now ask the player to add to it the next four numbers, $5(x + 2)$ (here, $252 + 253 + 254 + 255 + 256 = 1{,}270$);

3) ask for the result and from that you can recover the original number. All you have to do is divide by five and subtract 2, for $5(x + 2)/5 - 2 = x$ (here, $1{,}270/5 - 2 = 252$).

A shortcut in calculations

The world of numbers is vast and filled with possibilities. With a bit of inquisitiveness one can create games simply by devising new steps, or new ways to work out complicated and lengthy sums.

For one of these games you need two players. Ask each to write down a four-figure number on a piece of paper. Suppose the numbers are 1,223 and 1,887. One player (no matter which) is then asked to work out the product in the usual way. Meanwhile, you subtract 1,887 from 10,000, and 1 from 1,233, which gives $10{,}000 - 1{,}887 =$

8,113, 1,223 − 1 = 1,222. The second player is now asked to multiply these two numbers. Finally the two players are told to add their results. Before revealing them, however, you can announce that the sum is 12,221,887 (indeed, 2,307,801 + 9,914,086 = 12,221,887).

To clarify this let us examine the various steps algebraically. Let the two four-figure numbers be x and y.

1) The first player works out xy;

2) the second player works out $(10{,}000 - x)(y - 1) = 10{,}000y - 10{,}000 - xy + x$;

3) adding the two results yields $10{,}000y - 10{,}000 - xy + xy + x = 10{,}000(y - 1) + x$.

This final expression explains the trick: multiplying by 10,000 adds four zeros to the digits—for example, $13 \times 10{,}000 = 130{,}000$. Thus $y - 1$ gives the first four figures and x the remaining four. In our case $x = 1{,}887$ and $y - 1 = 1{,}222$, producing 12,221,887.

How much money is in your pocket?

Substituting a letter for a number—x, or any other letter—may seem almost elementary, but it was actually a major step in the development of mathematics, as it helped to illuminate the formal features of numbers and raised analysis to a more abstract level. When we see "652," we automatically think of a number. If, however, we see an algebraic expression, such as $10x + 9$, it is less clear that it too is a number.

We know that in algebra x can take any numerical value. If it equals 4, then the number just mentioned will be 49; if x equals 1, the number is 19, and so on. This gives rise to yet more games which may seem perplexing at first. In this example we see that the values of x appear in the tens of the answer, as is obvious if we take away the units. Consider another example.

We tell someone we can guess the amount of small change in his pocket if he will do the following:

1) Start with the total sum, s (say, 35 cents);

2) multiply by 2, $2s$ (here, $2 \times 35 = 70$);

3) add 3, $2s + 3$ (here, $70 + 3 = 73$);

4) multiply by 5, $5(2s + 3) = 10s + 15$ (here, $5 \times 73 = 365$);

5) subtract 6, $10s + 9$ (here, $365 - 6 = 359$).

We ask for the result, take away the units and are left with the sum of 35.

Other expressions too can generate this kind of trick, indeed a host of tricks. For example, use any number x and proceed as follows:

1) Take a number, x;

2) add 2, $x + 2$;

3) double, $2x + 4$;

4) subtract 2, $2x + 4 - 2 = 2x + 2$;

5) divide by two, $(2x + 2)/2 = x + 1$;

6) subtract the original number, $x + 1 - x = 1$.

The fifth step gives the vital clue: To get x we merely subtract 1.

How to guess a birth date

In the preceding algebraic expressions there was only one unknown and the trick was built around it. In the same fashion, we can devise tricks using expressions with two unknowns and find two numbers.

Consider an exercise that allows us to determine a person's birthday. First assign the numbers 1 to 12 to the months, starting with January. Let m be the month and d the number of the day we are seeking. Now put the person through these steps:

1) Multiply by 5 the number of the month, $5m$ (suppose the birthday is 13 June, then $5 \times 6 = 30$);

2) add 7, $5m + 7$ (here, $30 + 7 = 37$);

3) multiply by 4, $20m + 28$ (here, $4 \times 37 = 148$);

4) add 13, $20m + 41$ (here, $148 + 13 = 161$);

5) multiply by 5, $100m + 205$ (here, $5 \times 161 = 805$);

6) add the number of the day, $100m + 205 + d$ (here, $805 + 13 = 818$);

7) subtract 205, $100m + d$ (here, $818 - 205 = 613$).

Now ask for the number. The hundreds give the month, namely 6 for June, while the rest, 13, gives the day. Try a different one. Suppose we are to guess the date on which the Bastille fell (14 July 1789), marking the outbreak of the French Revolution.

1) Multiply the month by 5, $5m$ (here, $5 \times 7 = 35$);

2) subtract 3, $5m - 3$ (here, $35 - 3 = 32$);

3) double, $10m - 6$ (here, 64);

4) multiply by 10, $100m - 60$ (here, 640);

5) add the day, $100m - 60 + d$ (here, $640 + 14 = 654$).

Given this number, we now add 60, leaving $100m + d$ (here, 714). The values of d are found in the tens and units, while $100m$ is found in the hundreds, preventing the two from overlapping. Now we simply read off m and d. Of course, d must remain below 100, which limits the game to age, shoe size and so forth.

Guessing age and size of shoes

Here is an analogous game with a few confusions deliberately added. Suppose we are to guess the size of a person's shoes as well as his age. We proceed thus:

1) Multiply the number of years (a) by 20, $20a$ (if a is 20, we have $20 \times 20 = 400$);

2) add the number of the present day (d), $20a + d$ (here, supposing it is the 9th, $400 + 9 = 409$);

3) multiply by 5, $100a + 5d$ (here, $5 \times 409 = 2{,}045$);

4) add the shoe size (s), $100a + 5d + s$ (if s is 11, then $2{,}045 + 11 = 2{,}056$).

Now we subtract five times the number of the current day, which is known, leaving $100a + s$. The hundreds give the age and the rest gives the shoe size (here, $2{,}056 - 45 = 2{,}011$). The person is 20 and wears size 11 shoe.

Where is the error?

Behind every mathematical game there lies a wile. Many such games rely simply on people being unable to follow the various algebraic steps. However, we can invent very subtle tricks based on the procedure itself. Suppose we wish to prove that $1 = 2$. Take any two numbers x and y and suppose:

1) $x = y$;
2) multiply by y, $xy = x^2$;
3) subtract y^2, $xy - y^2 = x^2 - y^2$;
4) factorize, $y(x - y) = (x + y)(x - y)$;
5) divide by $(x - y)$, $y = x + y$;
6) by 1), $x = 2x$;
7) divide by x, $1 = 2$.

Each step seems correct; yet there is an error, an illogical step. When there is a contradiction in mathematics, the mistake can be found in the procedure or in the premises. If a game is irritating to play, it could be that either it is faulty (the premises are wrong), or the player is not sticking to the rules (the procedures). In this instance, the logical error lies in 5) when we divide by $(x - y)$, which because of 1) is zero. Clearly it makes no sense to divide by zero. We can now see that the contradiction was produced by introducing an error into the procedure.

In the following numerical expressions there are two mistakes for the reader to detect:

1) $2 + 1 - (-1) = 4$;
2) $6 \div \frac{1}{3} = 2$;
3) $(3 + \frac{1}{5})(3 + \frac{1}{8}) = 10$;
4) $18 - (-8) = 26$;
5) $-32 \times (27 - 27) = -32$.

This is the solution. The errors are in 2), where the result is 18 (dividing by $\frac{1}{3}$ is multiplying by 3); and in 5), where the result is zero (multiplying by zero, represented here by $27 - 27$, is zero).

Positional notation of numbers

This method (cf. p. 10) was used by the ancient Indians and was spread throughout medieval Europe by the Arabs. At the time it represented enormous progress in mathematics. Today we are so habituated to using Arabic numerals that we seldom realize the system's advantages. To do so we need only recall the Roman system which was long, cumbersome and a ready source of errors. Arabic figures are less intuitively obvious, but from the start they have exhibited a peculiar feature on which mathematical thinking rests: ever-increasing simplicity and generality. Take a number in Arabic figures, say 6,245. Here 6 indicates thousands, 62 hundreds, 624 tens and 6,245 units. We can write it also as $6(1{,}000) + 2(100) + 4(10) + 5(1)$, or $624(10) + 5(1)$.

Let us now consider a four-figure number algebraically, writing the digits as x_3, x_2, x_1, x_0. We can then write the number as $x_3(1000) + x_2(100) + x_1(10) + x_0(1)$. The first term indicates the figure with a positional value of thousands, the second hundreds, the third tens, and the fourth units. We can split this into $x_3(999 + 1) + x_2(99 + 1) + x_1(9 + 1) + x_0$, or, rearranging, $9(111x_3 + 11x_2 + x_1) + x_3 + x_2 + x_1 + x_0$.

Given that x_3 is the number of thousands, x_2 the number of hundreds, x_1 of tens and x_0 of units, consider the following set of instructions:

1) Take a four-figure number (say, 3,652);
2) write down the figure of thousands, $x_3 \cdot 1$ (here, 3);
3) write down the figure of hundreds, $x_3 \cdot 10 + x_2 \cdot 1$ (here, 36);
4) write down the figure of tens, $x_3 \cdot 100 + x_2 \cdot 10 + x_1 \cdot 1$ (here, 365);

5) add these, $x_3(111) + x_2(11) + x_1$ (here, $3 + 36 + 365 = 404$);

6) multiply by 9, $9(111x_3 + 11x_2 + x_1)$ (here, $9 \cdot 404 = 3{,}636$);

7) calculate the sum of the digits, $x_3 + x_2 + x_1 + x_0$ (here, $3 + 6 + 5 + 2 = 16$);

8) add to the previous sum, $x_3\ (1{,}000) + x_2(100) + x_1\ (10) + x_0$ (here, $3{,}636 + 16 = 3{,}652$).

We have just reconstituted the number through a new sequence of steps.

One rotten apple can spoil the whole basket

In the section "Where is the error?" we elicited paradoxical or contradictory results by dividing by zero in an algebraically unclear manner. Let us look at zero more closely in its various mathematical and philosophical senses. We know from school what the reciprocal of a number is. If the number is 6, its reciprocal is $\frac{1}{6}$; if it is 12, it is $\frac{1}{12}$, and so on. The larger the number, the smaller its reciprocal and conversely. Thus, in the sequence $\frac{1}{2}, \frac{1}{3}, \frac{1}{4}, \frac{1}{5}, \frac{1}{6} \ldots$ the terms become ever smaller. Using this method, one might imagine we could reach the smallest number in the world. Similarly, we might ask the meaning of dividing by zero, namely $\frac{1}{0}$. Is this a number at all?

Suppose someone discovered that $\frac{1}{0} = x$. Now in a normal case, say $\frac{28}{4} = 7$, we find $7 \times 4 = 28$. Therefore, in the case of $\frac{1}{0} = x$ we should find $x \cdot 0 = 1$; but $x \cdot 0 = 0$, otherwise we would have $1 = 0$. Thus there is a contradiction; the same would occur were we to divide by zero any number different from zero. There is one odd exception, namely dividing zero by zero; the result can be any number. For instance, take any number x, then $0 = 0 \cdot x$, and therefore $0/0 = x$. It is pointless, which is why division by zero is not allowed in mathematics.

All this is simply an intellectual game, rigorous, to be sure, but still a game. In mathematical thought the simplest steps can conceal quite profound concepts and principles. One of these principles belongs to logic—the science of correct reasoning—and states that from a contradiction any assertion can be established. Or, to quote medieval logicians: "ex absurdis sequitur quodlibet"—from the absurd anything follows.

Dividing a number by zero can produce contradiction, dividing zero by zero can yield any number. If we cancel by zero in $18 \times 0 = 3 \times 0$, we get the contradiction $18 = 3$. Or, as in the following: if $x = 1$, then $x^2 - x = x^2 - 1$, $x\,(x - 1) = (x + 1)\,(x - 1)$, and cancelling by $(x - 1)$ gives $x = x + 1$, $1 = 2$; we have divided by $x - 1$, which in this case equals zero. It is an error we frequently make in mathematics; even Einstein once inadvertently did so.

The symbol 0 (zero) came to the West with the Indo-Arabic numerals. It is one of the most useful symbols, but also one of the most ambiguous and contradictory. Like the other numerals, zero has a positional meaning. In 432 the 2 does not have the same meaning it has in 423. This is also true in 430 and 403 where the zero means the absence of units in one case and the absence of tens in the other. The concept of zero has been developed further in mathematics, and to an even greater extent in philosophy and religious thought. If we think of zero, we think of nothingness, but what is that? Roughly we might say that nothingness is the denial of existence, it is that which is not. Yet, as we think of nothingness, it must somehow exist. In short, we have an unfathomable concept and that creates paradoxes.

Originally the notion of nothingness was extraneous to Greek philosophy as the Greeks would not accept the being of that which does not exist. Indeed, zero does not enter the Greek and Roman numeral systems. It is probably the philosopher Zeno (336–264 B.C.), a Phoenician from Cyprus and founder of Stoicism, who introduced this non-Greek concept into ancient philosophy.

Ordinary language and mathematical language

Many of us tend to think of mathematics as simply a practical tool for accounting and measuring. At best, we have some smattering of science, and are familiar with a set of techniques and methods of mathematical analysis that make our calculations work. However, we fail to grasp a basic feature of mathematics, namely its language.

The term "language" suggests everyday language which conveys information. However, language has other tasks, such as organizing our cognitive activities to clarify our concepts and to represent our results. Mathematics, with its abstract symbols, fulfills this function very well. Still, too often mental habits, learned in school in mechanical ways and devoid of mathematical insight, make us see mathematics as containing a different rationality and as something apart from everyday language. This is not only absurd but artificial. It is even more absurd to assume there are two languages representing opposing ways of facing reality. It is true that mathematical language is particularly appropriate for descriptions of certain problems and their solutions, but this does not justify divisions in knowledge.

Number games, especially algebra games, force us to connect everyday language and mathematical language and to translate them back and forth. This proves there are no genuinely isolated areas, even if everyday language is more complex and varied and therefore better suited to recounting subjective and personal situations. Mathematical symbols and the relations between them are abstract, synthetic mental constructs

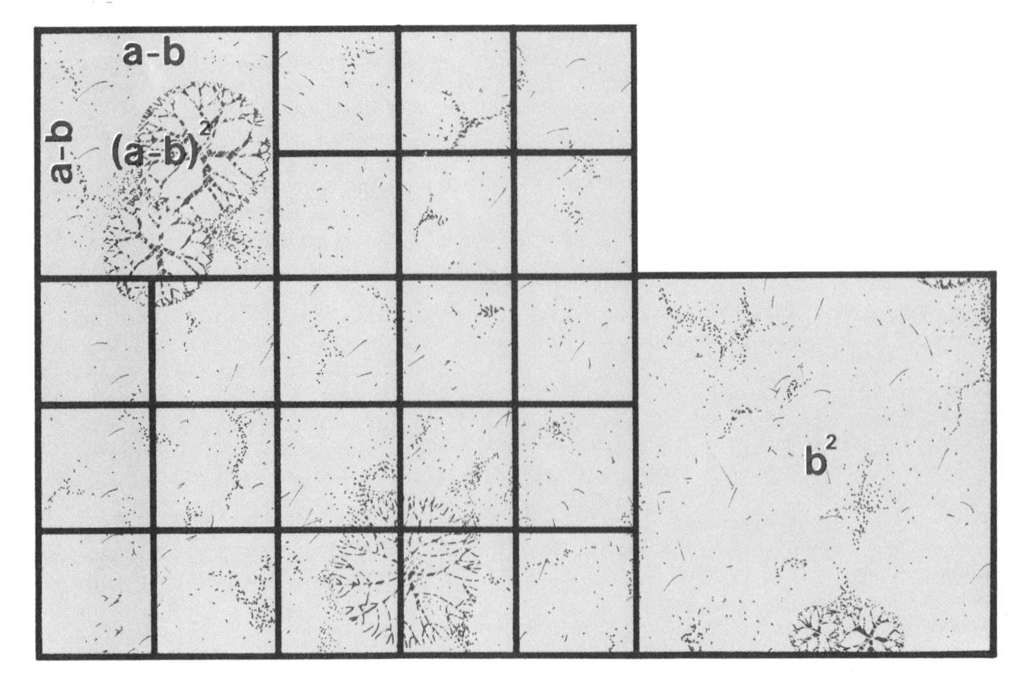

$$(a-b)^2 = a^2 + b^2 - 2ab$$

Too often we think of the rules of algebra as simply abstract, and we forget that mathematics began for very practical and concrete reasons. Take the product of two binomials: this is easier to understand if it is linked with the problem of dividing land.

Left: A geometric illustration of the simple algebraic problem of squaring $(a - b)$. We know from algebra that this is $(a - b)^2 = a^2 + b^2 - 2ab$. In the diagram, we find it is the small square at the top left.

and unlike ordinary language, they are specific and unambiguous. Translating from mathematical into everyday language is therefore a particularly useful exercise, especially at school age. Take the expression $x/3 + 5 = x/2 + 6$, for example. This equation and its solution can help us to formulate a problem first in everyday language, and then synthetically in the language of mathematics. In ordinary language the equation is translated: *A third of a number increased by five equals half that number increased by six, if that number is minus six;* mathematically, $x/3 + 5 = x/2 + 6$, yields $x = -6$. Similarly, the expression $(x - 2)/4 = (5 - x)/6$ can be read as: *A quarter of two less than a number equals one-sixth of the difference between five and that number.*

Let the reader try to translate the expression $AB = AC$. If AB and AC are two segments, we say that the segment AB is congruent with (equal to) the segment AC. If ABC is a triangle, $AB = AC$ tells us that the triangle with vertices at A, B, C is isosceles. If BC is a segment, we can say that A is the middle point (Figs. 5, 6, 7).

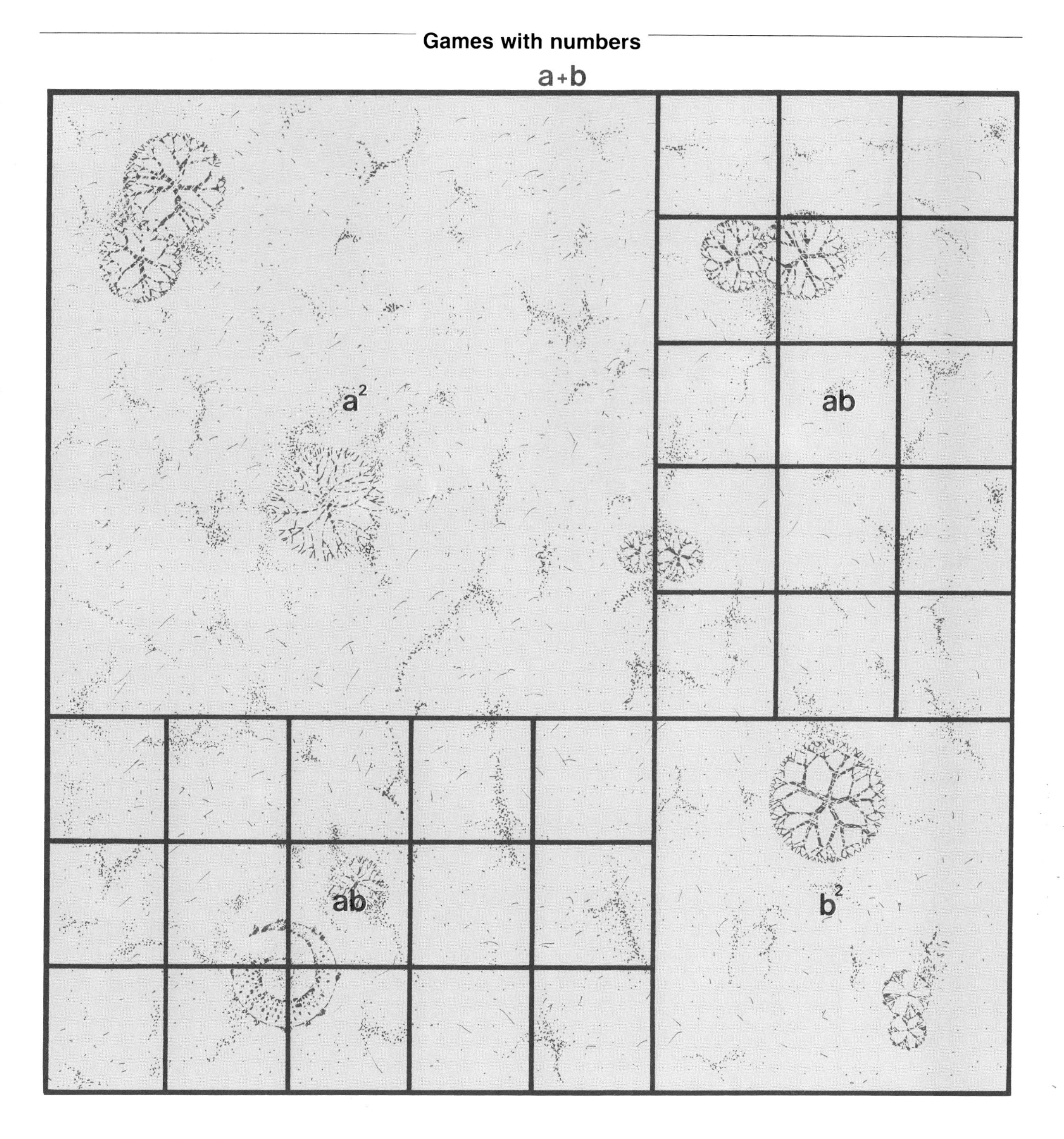

$$(a+b)^2 = a^2 + b^2 + 2ab$$

In this diagram the connection between an algebraic expression and a geometric one is more immediate. When we multiply $(a + b)$ with itself, we get the squares of each term plus twice their product: $(a + b)^2 = a^2 + b^2 + 2ab$.

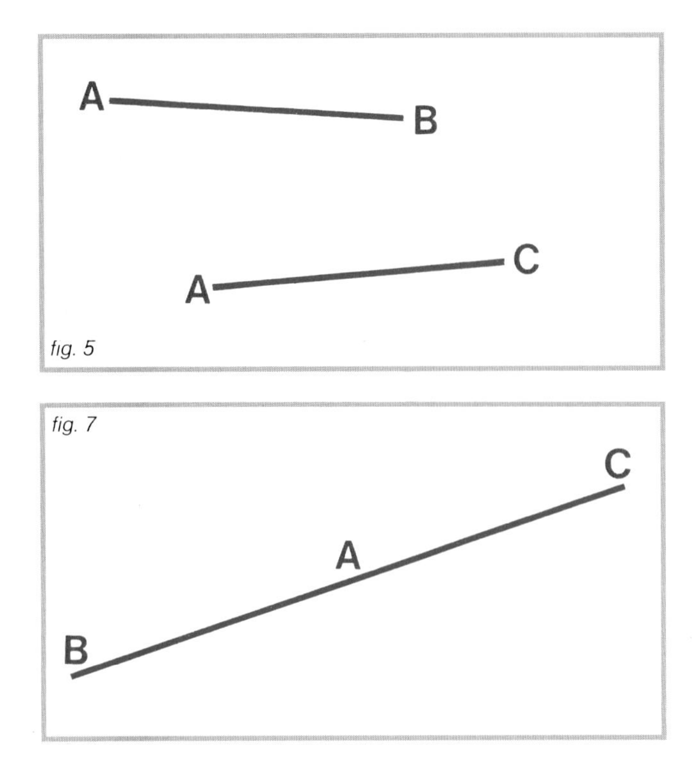

fig. 5

fig. 7

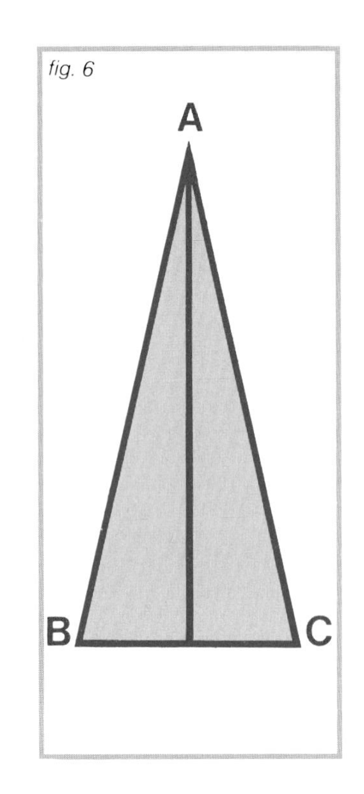

fig. 6

GAMES WITH GEOMETRICAL FIGURES

God geometrizes continually (Plato)

Geometry and optical illusions

In Fig. 8-9, which is longer, *AB* or *CD*? In fact neither. They are equal. Now take Fig. 10. *CD* looks longer than *AB*, but again they are equal, as a ruler will show. These are optical illusions. Through our sense organs we perceive vital information about our surroundings. That data then travels to the brain where it is processed and sent to us as sense experience. There are visual, auditory, gustatory, olfactory and tactile sensations, depending on the sense involved. However, our senses can deceive us and give us an incomplete picture of reality. Moreover, our senses can be conditioned by previous experiences or habit and thus create illusory sensations. Look at Fig. 11. You probably notice a triangle. None is drawn, but the more one looks the more one seems to perceive a triangle, even if reason tells us the drawing consists of circular sectors.

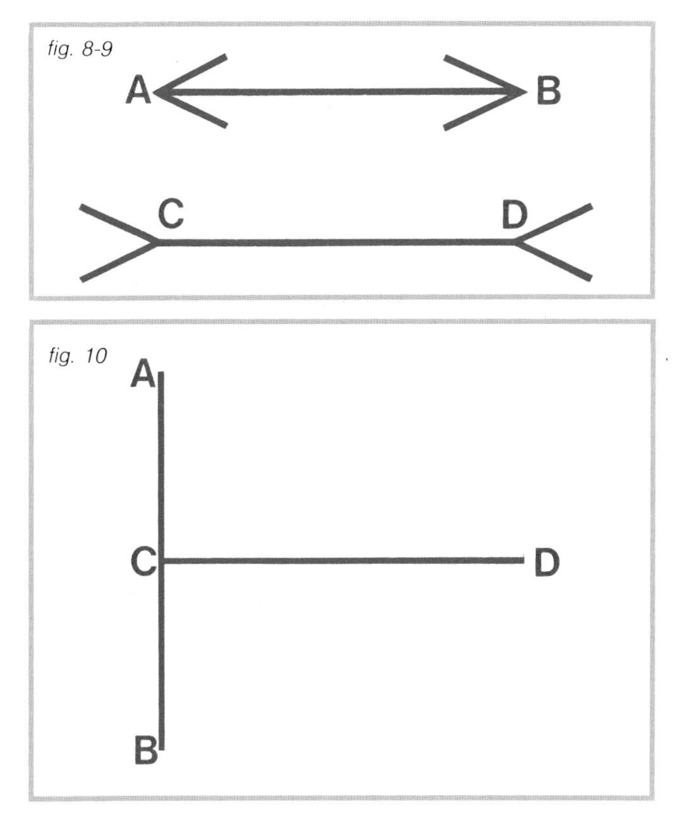

(continued on page 38)

Below: This print, by the Dutch artist Maurits Cornelius Escher (1898–1972), *Ascent and Descent*, was inspired by the ambiguity in certain geometrical figures. The monks in the outer row seem to ascend endlessly while those in the inner row seem to descend endlessly. (© Beeldrecht, Amsterdam 1982)

Right: The steps on which Escher's optical illusion is based. In three dimensions we cannot represent or build a staircase that ends where it begins. In a two-dimensional picture Escher overcame these limitations by altering certain figurative signals and visual data.

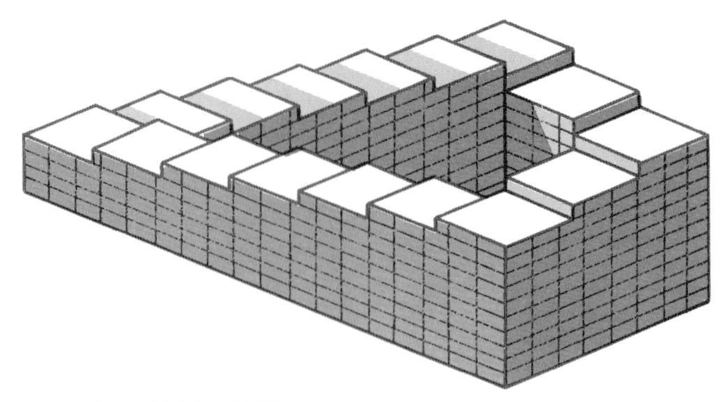

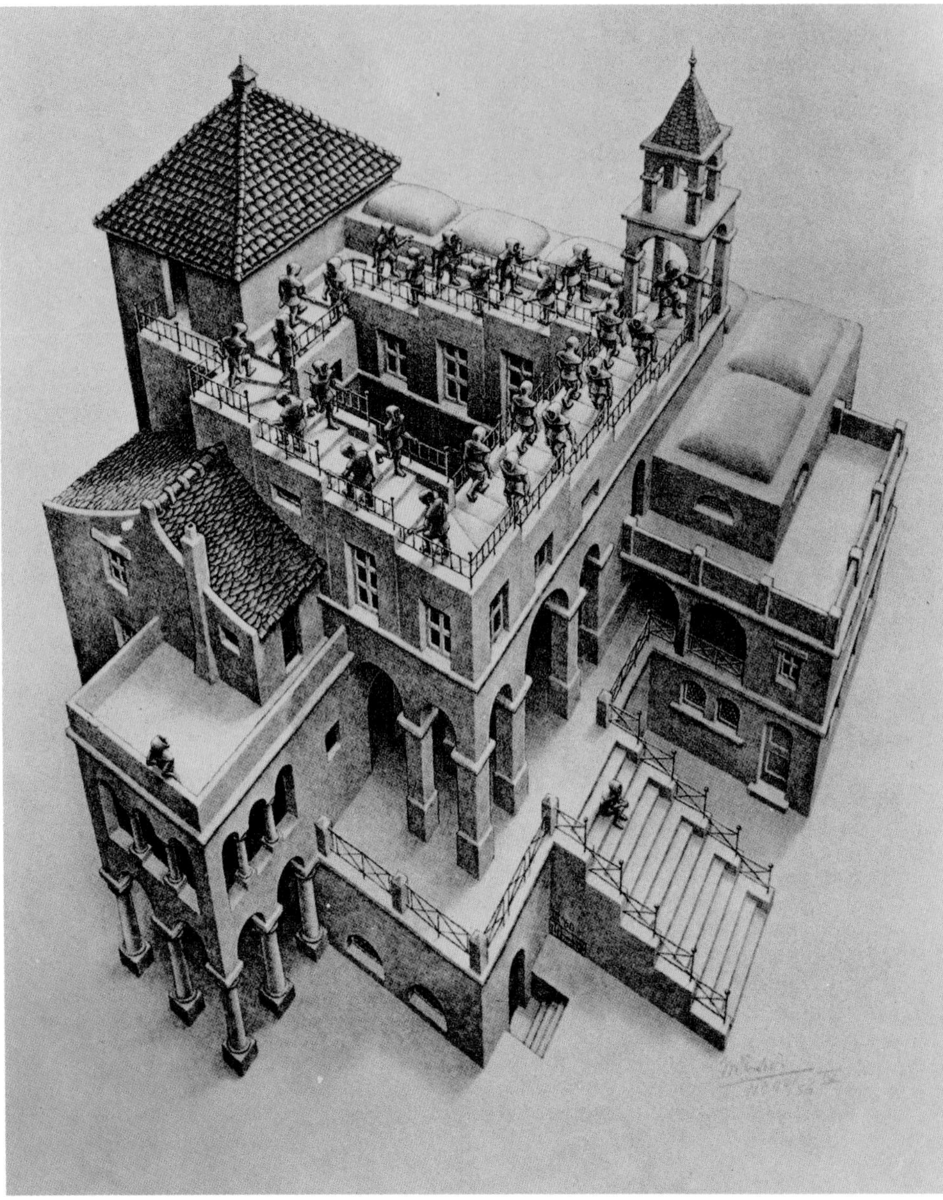

Opposite: Bridget Riley's, *Cataract III*, 1967, London. (British Council, by permission of Rowan Gallery.) This Op Art (Optical Art) painting plays on optical illusion. Such phenomena warn us that our perception of an image can be different from its reality, which is perhaps what attracts our attention.

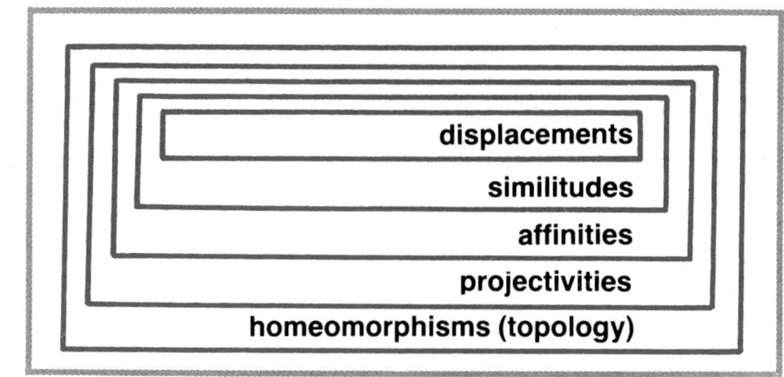

Opposite: Regular similar geometric figures converge to the center of the dome of Sheik Loftollah's mosque at Isfahan (Iran). This is a typical example of Arabic art, in which geometrical themes often occur. In the Arabic world, the abstract sciences, particularly mathematics and geometry, were based on religion. In Islam, Allah represents life itself. Man, or plants or animals are only single parts of the universe and no one of them can represent the totality. Indeed Arabic artists seldom portrayed people or animals, choosing instead to create abstract geometric patterns in which animals were seen only in stylized forms.
Above: Until about the second half of the 19th century the geometry in use was basically Euclidean, namely the intuitive geometry we learn at school. Euclidean geometry was regarded as an immutable and predetermined way to grasp phenomena and experience, witness the philosopher Immanuel Kant. However, in the second half of the 19th century some mathematicians and scientists (Gauss, Lobachevsky, Bolyai) discovered that other geometries could be constructed, leading to non-Euclidean spaces, simply by denying one or more of Euclid's basic principles. In 1872, the mathematician Felix Klein at Erlangen proposed a geometrical research program to radically change figurative geometry into a system of transformations. This means there are many geometries, and together they form a system in which each can be constructed from the simplest to the most general. Euclidean geometry is a metric geometry and belongs to the group of displacements; it allows only isometric transformations (displacements, rotations, symmetries) that vary the position but not the size of angles and lines. If lines are allowed to change in size, we have the group of similitudes; if angles too can change, we have affine geometry; if parallels change, we have projective geometry. Finally, a figure may be continuously distorted as long as connected parts are not severed and points are not superimposed; this is the group of homeomorphisms or topology. Each of these geometries is weaker than its predecessor but broader as what remains are the more general properties of figures. The way the exact sciences developed at the time (Einstein's relativity, quantum theory) confirms that the new ideas on space constructed for non-Euclidean geometries were more in accord with reality than the absolute space of classical physics.
Right: The perspective of an Arab portico, a clear example of projectivity.

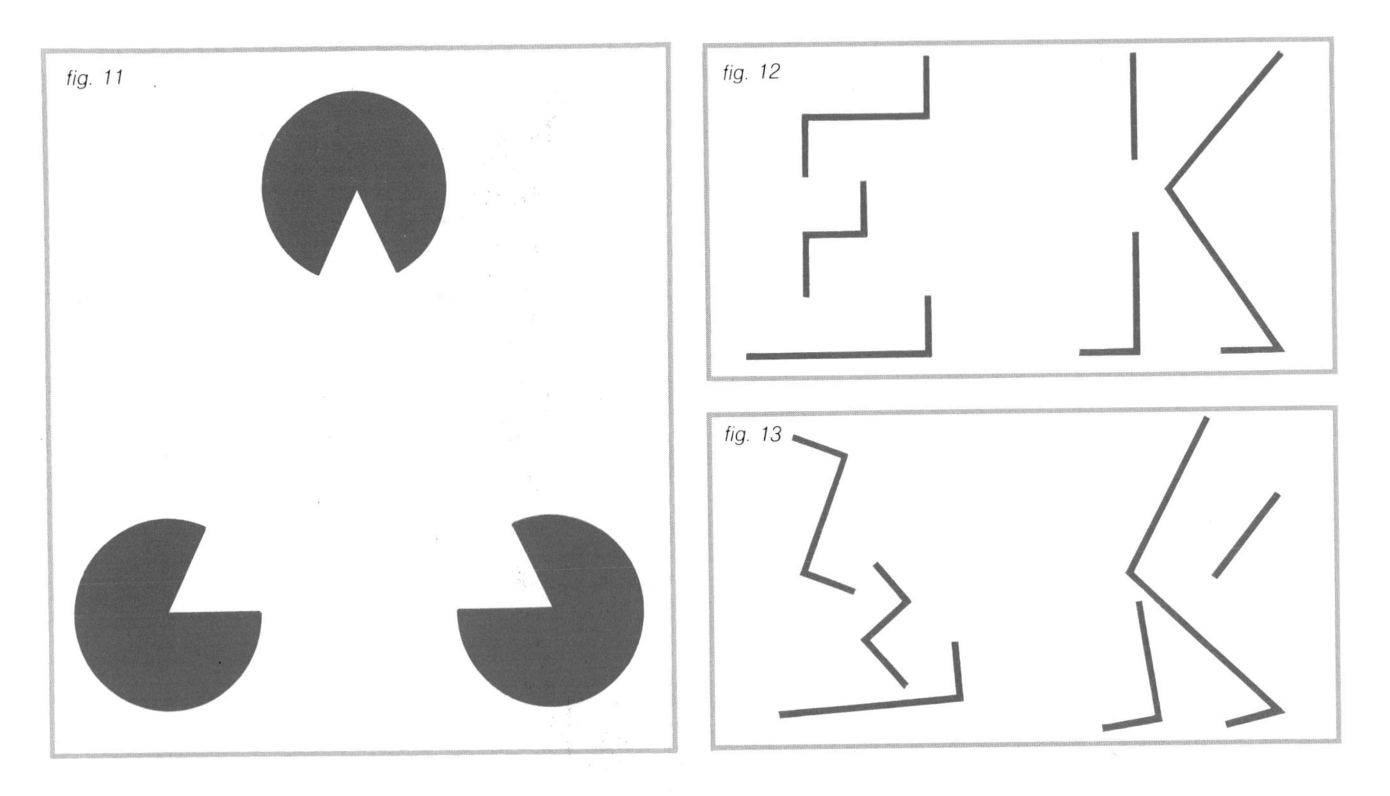
fig. 11

fig. 12

fig. 13

Many factors intervene in the processing of data supplied by the senses; particularly important are our perceptual habits. The mechanism that makes us see a triangle in Fig. 11 is the same one that makes us see two letters in Fig. 12 but not in Fig. 13.

Consider a curious experience we have all had at the movies. In a Western, for instance, as we watch the pursuit of a coach moving faster and faster, suddenly the coach's wheel spokes seem to rotate backwards. Obviously our sensations are confused; our eyes tell us the coach moves forward while the wheels suggest that it is going backward. When the chase ends, the wheels just as suddenly resume their proper forward rotation. Our senses have deceived us and we have experienced an optical illusion. The explanation is quite simple and is based on the physical principles of cinematography. The sensation of movement we perceive in the film is caused by the fact that the slightly different individual frames that are projected on the screen at the rate of 20–24 a second, "persist" for an instant on the eye's retina in an ordered sequence, hence the sensation of motion.

Why then do we see the spokes move backward after a certain speed is reached? We can only give a general indication, as a detailed account involves the psychology of perception and principles of optics too complex to outline briefly. Roughly, this phenomenon occurs because the time a spoke takes to move from one position to the next steadily decreases, until it is less than the appearance of an individual frame on the screen. As a result, the pictures on our retina merge, giving the impression of a reversal of rotation. This can be seen more or less in Fig. 14. The fourth diagram shows the reversal point. Note that the distance between the two spokes increases.

In discussing this common optical illusion we have mentioned some physical principles and mechanisms of perception. Optical illusions are

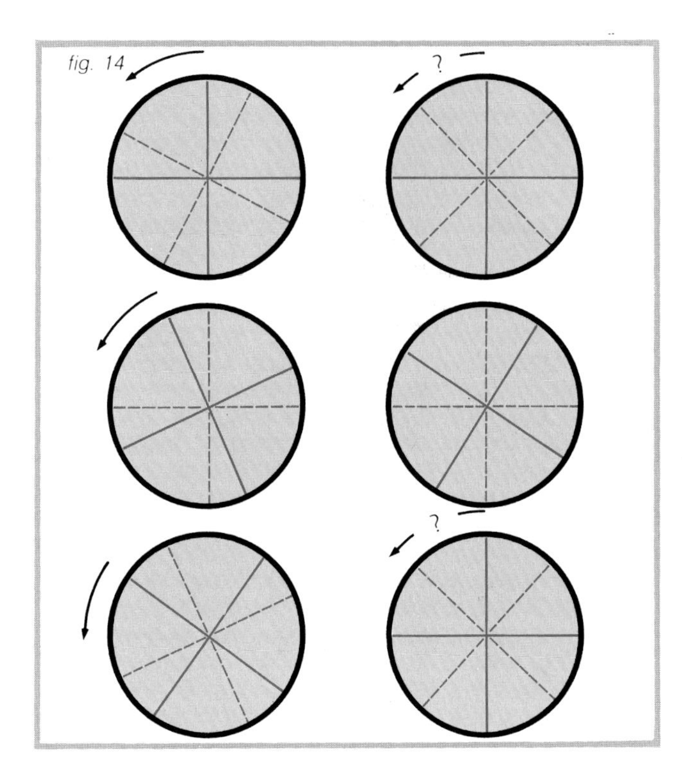

fig. 14

fig. 16

indeed particularly important in many fields: art, psychology, mathematics, even philosophy. Philosophers ask to what extent our senses give us information that correctly reflects the world around us and what subjective elements distort our perception of reality. In any event, optical illusions have decisively influenced the psychology of perception; they are a popular instrument for studying how the brain organizes and interprets what the senses convey to it.

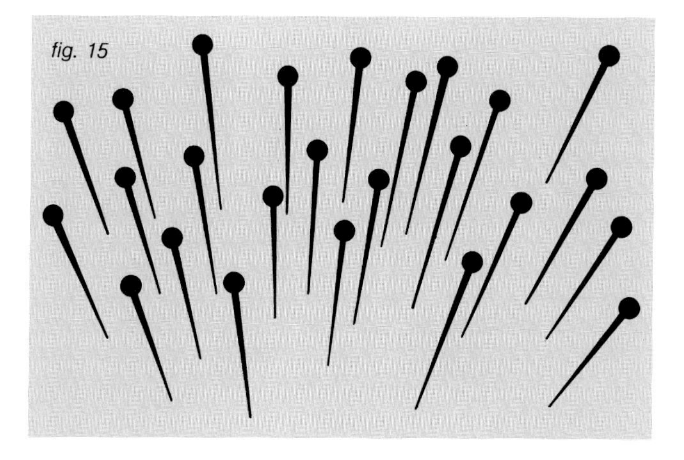
fig. 15

Look at the pins with their heads up in Fig. 15. Now raise the book to eye level, hold it horizontally and close one eye. The pins seem to be standing up. Among the illusions studied and analyzed by Gestalt psychology (the psychology of the perception of shape, adopted by the Germans Max Wertheimer, Kurt Koffka, Wolfgang Köhler, Kurt Levin among others) are those concerning figures that can be perceived in two equally valid ways.

Fig. 16 shows the new flag adopted by the Canadian House of Commons in 1965. The middle panel has a maple leaf on it, but if we concentrate on the white background surrounding the leaf, we seem to perceive two angry faces. Such figures are called "reversible," because their mental representation can suddenly switch without any change in the visual information to the eye.

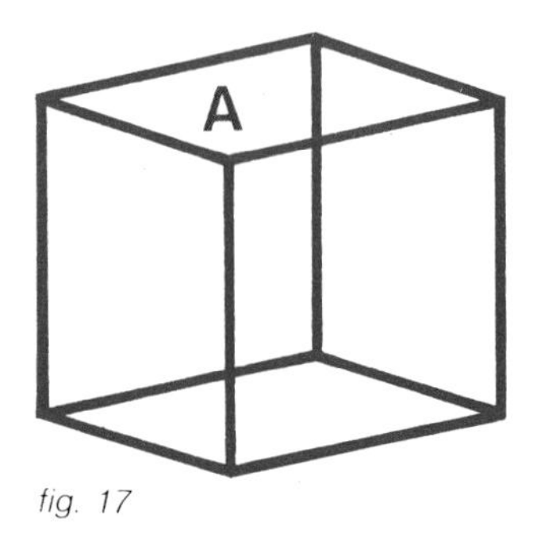

fig. 17

fig. 19

A well-known figure of this type is "Necker's cube," or transparent cube. In 1832 Louis Albert Necker, a Swiss geologist, observed the "perspective inversion" and reported that some drawings of transparent rhomboidal crystals present two different pictures in which front and back are interchanged.*

In the cube of Fig. 17, one side seems to be in front, but when you look at the cube steadily the sense of depth reverses and the side at the back suddenly appears in front. To convince yourself, look at the corner A and watch how it jumps from back to front.

How do we explain this? Reversible figures produce a set of data that can be given two equally valid interpretations; the brain accepts one first and then the other. Another classic example is the goblet of Edgar Rubin (a Danish psychologist). In Fig. 18, initially we see two faces, then a goblet.

The inversion of object and background is only one reason for optical illusions; other elements in a figure can also produce ambiguity. Take the

*A rhomboid is a parallelogram in which the angles are oblique and adjacent sides are unequal.

fig. 18

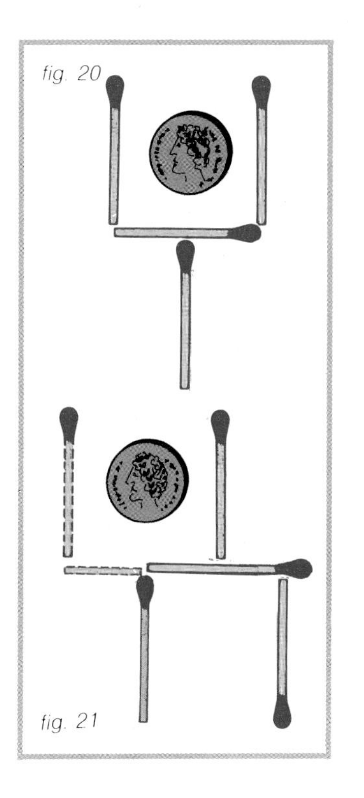
fig. 20
fig. 21

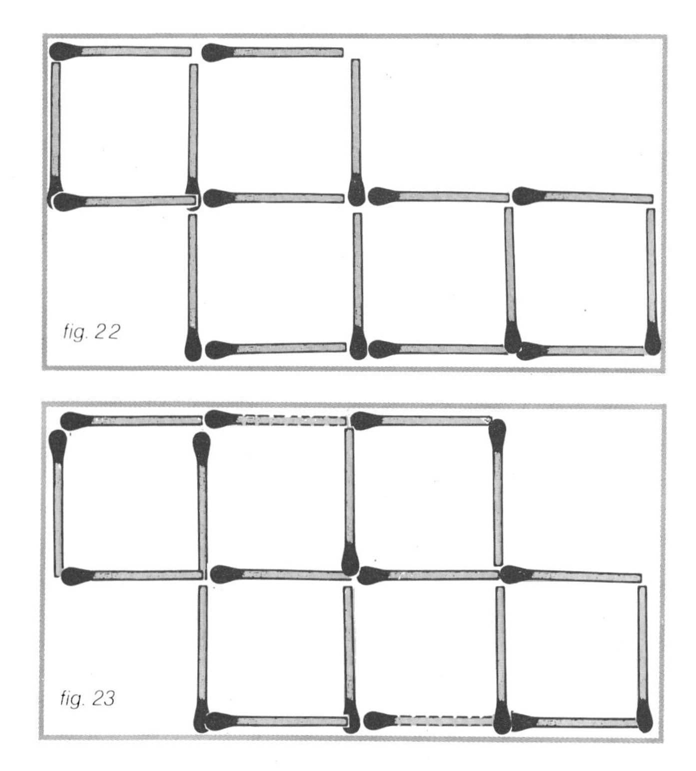
fig. 22
fig. 23

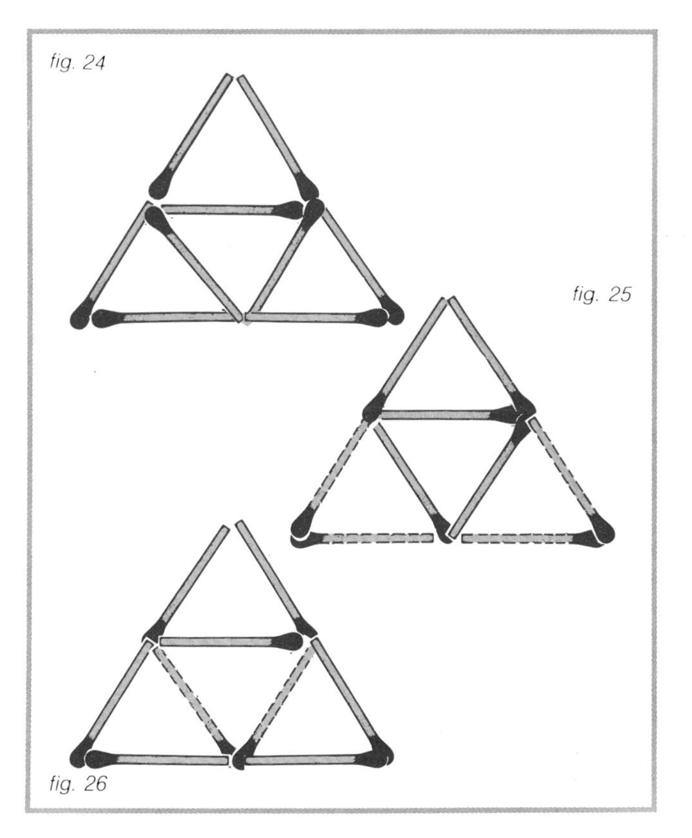
fig. 24
fig. 25
fig. 26

picture in Fig. 19. It looks like a duck or a rabbit, depending on whether we focus first on the left side or on the right side. Indeed, the duck-rabbit was devised by psychologist Joseph Jastrow in 1900 to illustrate ambiguity.

Games with matches

The simplest figure games can be played just about anywhere. All we need is a box of matches. Fig. 20 shows a coin inside a chalice formed by four matches. By moving only two matches, how can we reconstruct the figure to put the coin on the outside? Fig. 21 shows the solution; the dotted lines indicate the initial position. The next figure illustrates a similar trick. Fig. 22 shows five squares formed by a certain number of matches. The problem is to remove one square by changing the position of only two matches (no open-sided or incomplete squares are allowed). One solution appears in Fig. 23.

Next, look at the triangle in Fig. 24. By removing four matches we always finish with two equilateral

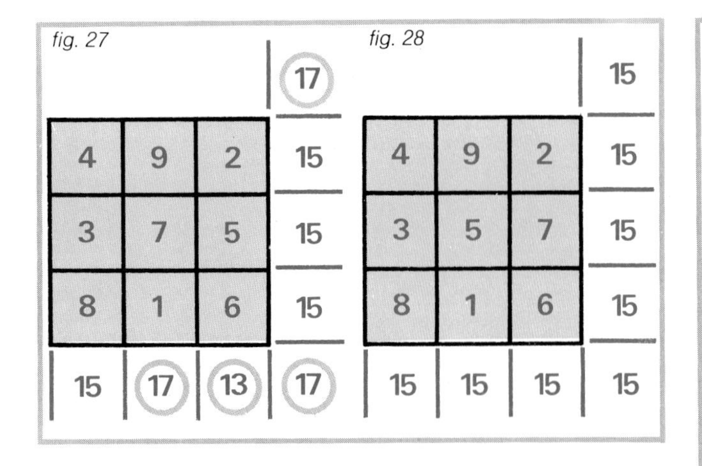

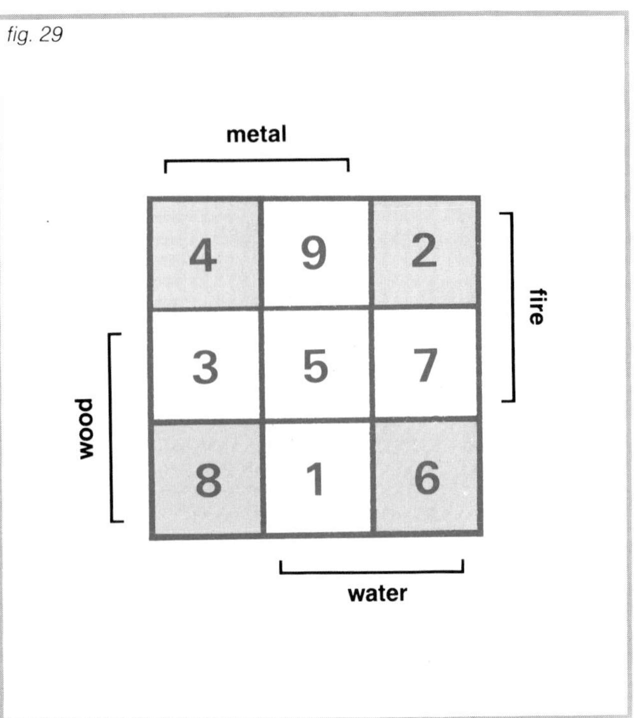

triangles (Fig. 25). Now try to produce two triangles by taking away only two matches (the figures must have no open sides). One solution is shown in Fig. 26.

Lo shu, an ancient Chinese figure

Take a square of nine equal boxes and, using numbers one through nine, write a number in each box in such a way that each row and column adds up to the same sum. This is quite an absorbing and difficult task and will probably be solved only by trial and error. One attempt, shown in Fig. 27, fails because the totals of the second and third columns and the diagonals are different from the totals of the rows and the first column. However, by interchanging 5 and 7 we reach the arrangement of Fig. 28, which is a solution. A figure such as this with rows, columns, and diagonals adding up to the same sum, is known as a "magic square." The Chinese were the first to discover the properties of magic squares which they called *Lo shu.* Legend tells us that the figure was revealed to man on the shell of a mysterious tortoise which crawled out of the river Lo many centuries before Christ. Historically, *Lo shu* goes back no further than the 4th century B.C. The Chinese attributed mystical significance to the mathematical properties of the magic square and made it a symbol uniting the first principles that shaped all things, man, and the universe, and that remain a part of them eternally. Thus, even numbers came to symbolize the female-passive or *yin* and odd numbers the male-active or *yang*. At the center on the two diagonals is the number five, representing the Earth. Around it are four of the major elements: metals, symbolized by four and nine; fire, by two and seven; water, by one and six; and wood, by three and eight. As Fig. 29 indicates, each element contains measures of both *yin* and *yang*, female and male—the opposites in reality.

Magic squares: their history and mathematical features

A magic square exhibits the integers from 1 to n^2, without repetition, in such a way that each row (left to right) and column (top to bottom) and the two

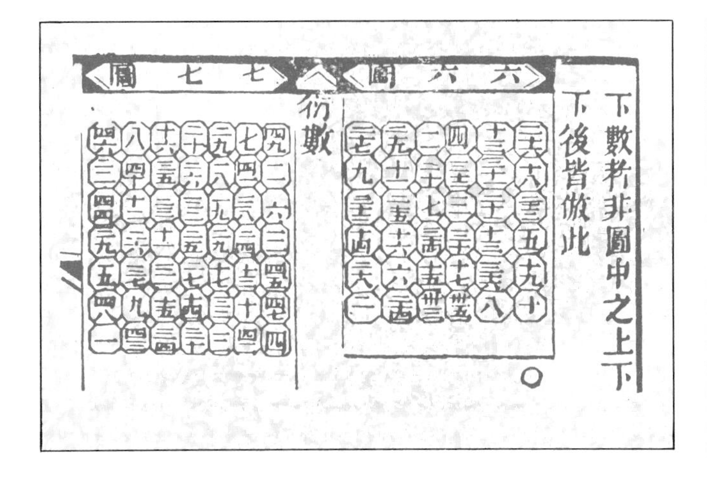

Above: A Chinese magic square of order 6, dating from about A.D. 1590. Chinese mathematicians were among the early students of the numerical and geometrical features of such squares. The Lo shu is a third-order magic square (Fig. 29) known since the 4th century B.C. and accorded special religious significance.
Right: The same square in Arabic numerals. All the columns, rows and diagonals add up to 111.

27	29	2	4	13	36	111
9	11	20	22	31	18	111
32	25	7	3	21	23	111
14	16	34	30	12	5	111
28	6	15	17	26	19	111
1	24	33	35	8	10	111
111	111	111	111	111	111	111

diagonals have the same sum. The number n is called the order, base, module, or root of the square.

The square's mathematical properties have captured the imagination of scientists since the origins of arithmetic and geometry; many of the ancient scientists attributed magical and cabalistic virtues to the square. Magic squares were known in India, and from there came to the West, probably through the Arabs. During the Renaissance, a period of extensive inquiry, the mathematician Cornelius Agrippa (1486–1535) worked on magic squares of orders greater than 2. Order 1 is trivial, and it can be proven that there is no magic square of order 2. (If in doubt, try to construct one.) From this Agrippa deduced that the ancient Greek philosophy incorporating four elements (fire, water, air and earth) as first principles was inadequate.

Agrippa constructed magic squares of orders 3 to 9 inclusive and gave them an astronomical meaning. They symbolized the seven planets known at the time (Saturn, Jupiter, Mars, Sun, Venus, Mercury, Moon). Copies of magic squares cut in wood or other materials once served as amulets and to this day are used in some parts of the East. In the 16th and 17th centuries people believed a magic square engraved on a small silver tablet could protect them from the plague.

The aura of magic surrounding these squares is partly based on the surprising number of combinations that can be fashioned with them. Consider the Lo shu, the oldest and simplest of magic squares. Is there a method for constructing a magic square? First, we list the eight ways in

$$9 + 5 + 1 = 15$$
$$9 + 4 + 2 = 15$$
$$8 + 6 + 1 = 15$$
$$8 + 5 + 2 = 15$$
$$8 + 4 + 3 = 15$$
$$7 + 6 + 2 = 15$$
$$7 + 5 + 3 = 15$$
$$6 + 5 + 4 = 15$$

fig. 30

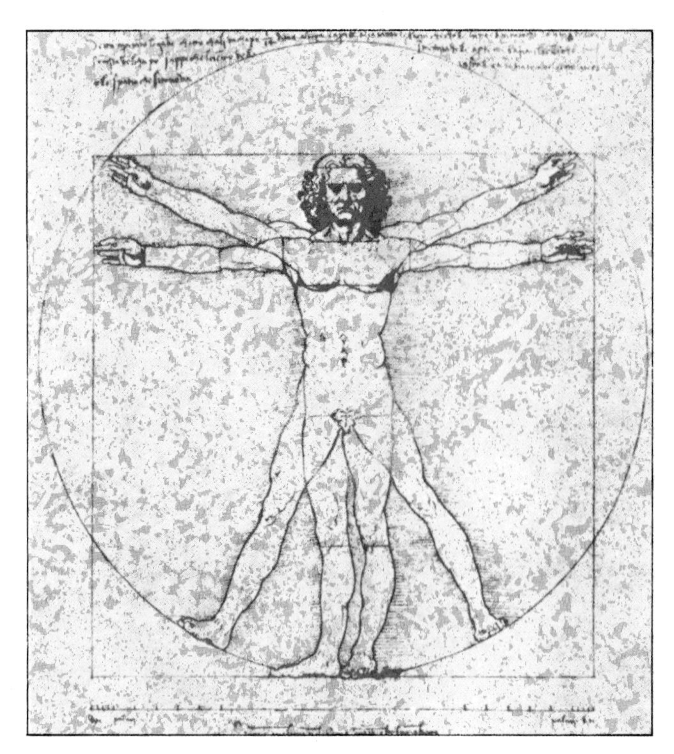

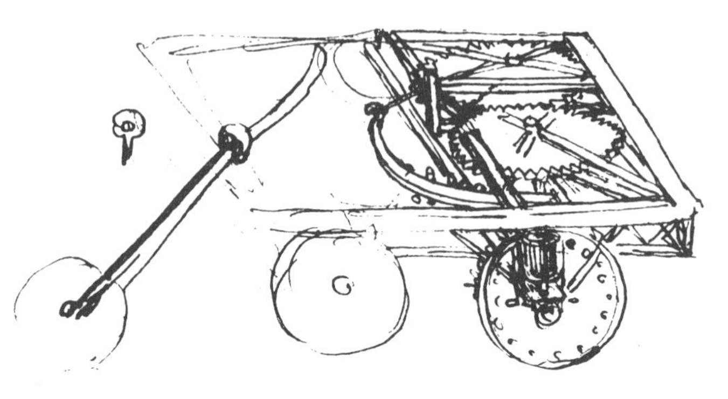

Left: Leonardo da Vinci's, *Canone delle proporzioni del corpo umano:* Rule for the proportions of the human figure, according to Vitruvius. Gallerie dell'Accademia, Venice.
Below: Leonardo's sketch for a self-moving vehicle powered by two large springs and steered by a handle fixed to a small rear wheel.

Opposite: Albrecht Dürer's, *Melencolia* (engraving, 1514). During the Renaissance, with the renewal of interest in the arts and sciences, two artists to show particular fascination with mathematics and geometry were Leonardo and Dürer. This picture reveals a magic square of order 4, to symbolize melancholy, then considered an energizing condition. The last line of the square indicates the year of composition, 1514.

which the number 15 can be produced from the first nine natural numbers (Fig. 30). Note that each number appears only once. Therefore we must arrange the triplets of equal sum (in Fig. 30, the first two, for example, share the nine) in such a way that two of them have one number in common. This is possible because the square has eight lines (three rows, three columns, two diagonals) that must add up to 15, and these correspond to the eight possibilities of Fig. 30. To fill the compartments, remember that each number can occur only once. Take the central compartment: It must appear in four triplets (one row, one column, two diagonals). The only such number is 5. Next consider the corners: Each must contain a number that appears in three triplets (one row, one column, one diagonal). Fig. 30 indicates there are only four such numbers: 8, 6, 4, 2.

We could have reasoned the other way around and counted how many times a number appeared in the set of triplets, and then deduced its position in the square. For example, 9 occurs only twice, and therefore cannot be at the center or in a corner. Given the square's symmetry, there can be two cases—mirror images of each other.

More intricate magic squares

The Renaissance was a period of cultural and artistic revival throughout 15th and 16th century Europe and touched all branches of learning. The development of mathematics and geometry was remarkable and, indeed, their influence extended to the figurative and architectonic arts for which they became a model and reference point. All these trends came together in the person of Leonardo da Vinci (1452–1519), scientist, writer, man of letters, engineer, mathematician and artist. For da Vinci, mathematics and geometry were closely linked to mans' artistic and cultural endeavors. In a no longer extant treatise on painting, *De pictura,* he wrote: "Do not read me if you are no mathematician." Actually Plato (428–348 B.C.) made the connection long before da Vinci and even saw fit to place this warning over the en-

MELENCOLIA I
16 3 2 13
5 10 11 8
9 6 7 12
4 15 14 1

fig. 31

16	3	2	13
5	10	11	8
9	6	7	12
4	15	14	1

fig. 32

2 × 2 = 4	2×9 = 18	2×4 = 8	2 × 15 = 30
2 × 7 = 14	2 × 5 = 10	2 × 3 = 6	2 × 15 = 30
2 × 6 = 12	2 × 1 = 2	2 × 8 = 16	2 × 15 = 30
2 × 15 = 30	2 × 15 = 30	2 × 15 = 30	2 × 15 = 30

trance to his school: "Let no one enter who masters no geometry."

The link between mathematics, geometry and art also underlies the work of the German painter Albrecht Dürer, a contemporary of da Vinci. In Dürer's noted engraving *Melencolia,* there is a magic square, often considered the first example of one seen in the West. It is constructed so the rows, columns and diagonals add up to 34 (Fig. 31). Moreover, the four central compartments add up to 34. The second and third compartments in the bottom row indicate the date of composition: 1514. Aside from the intimacy between the arts and sciences during the Renaissance, perhaps another reason for Dürer's inclusion of a magic square in his engraving, is that fourth-order squares were thought to possess special therapeutic virtues. Indeed, astrologers of the period advised wearing them as amulets to dispel melancholy.

Let us move on to other magic squares. Take the third-order square of Fig. 29 and multiply its entries by a constant to produce new entries. These will again add up to a constant. Indeed,

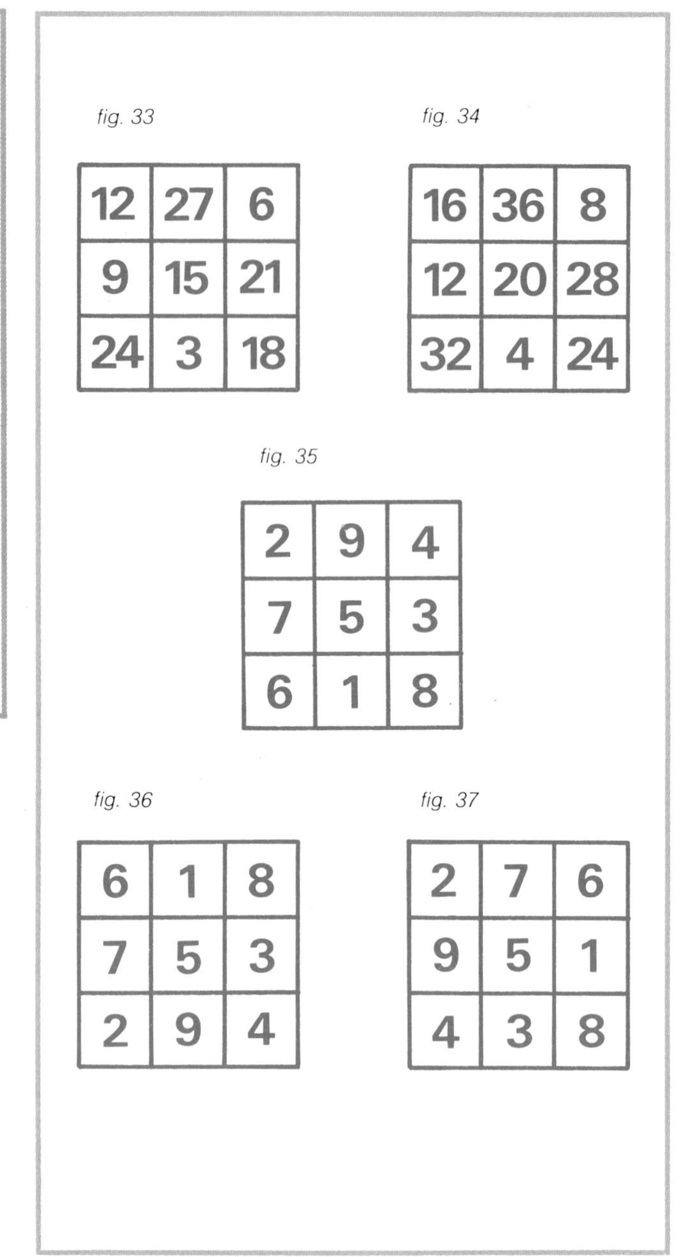

fig. 33

12	27	6
9	15	21
24	3	18

fig. 34

16	36	8
12	20	28
32	4	24

fig. 35

2	9	4
7	5	3
6	1	8

fig. 36

6	1	8
7	5	3
2	9	4

fig. 37

2	7	6
9	5	1
4	3	8

multiplying by 2 gives a sum of $2 \times 15 = 30$ (Fig. 32). Similarly, Figs. 33 and 34 are constructed by multiplying by 3 and 4 the entries of the square in Fig. 29. If we exclude rotations and reflections, there exists a unique magic square of the third order. Figs. 35 and 36 are only a single third-order square with a reflection about the cen-

16	2	3	13
5	11	10	8
9	7	6	12
4	14	15	1

fig. 38

7	12	1	14
2	13	8	11
16	3	10	5
9	6	15	4

fig. 39

4	15	6	9
5	10	3	16
11	8	13	2
14	1	12	7

fig. 40

7	12	1	14
2	13	8	11
16	3	10	5
9	6	15	4

fig. 41

b	c	a	
c	a		b
a		b	c
	b	c	a

fig. 42

	d	f	e
e		d	f
f	e		d
d	f	e	

fig. 43

tral row, while Fig. 37 has been rotated around a diagonal. With higher orders, the number of arrangements increases. A fourth-order square allows 880 different placings of its 16 numbers, excluding reflections and rotations. This was first discovered by the mathematician Bernard Frénicle de Bessy in 1693. Figs. 38, 39 and 40 show some of these solutions, with the sum 34. It is not yet clear what mathematical law governs the disposition of numbers in magic squares. The question remains open, and the known solutions have only been discovered by trial and error.

How many fifth-order magic squares are there? Until recently the estimate was about 13,000,000. In 1973, however, Richard Schroeppel, a programmer with Information International, determined the exact number with the aid of a modern computer (His findings were later published in *Scientific American,* vol. 234, no. 1, Jan 1976). Without counting rotations and reflections, there are 275,305,224 different solutions.

Diabolic squares

These are even more intriguing than magic squares, because of their additional properties. Again consider Dürer's square as it is rearranged in Fig. 39 and repeated in Fig. 41. The sum of the four central squares is $13 + 8 + 3 + 10 = 34$, as is the sum of the four corners and the vertical as well as the horizontal off-diagonal squares. Such magic squares are called "pandiagonal." The same constant results if we add the set of four numbers marked in Figs. 42 and 43.

Similarly we can form fifth-order magic squares with particular properties. In Fig. 44 the sum is always 65. Considering corner numbers plus the central one, as well as diagonal numbers (including the central one), we always produce the same sum: $9 + 13 + 17 + 1 + 25 = 65$, $7 + 5 + 21 + 19 + 13 = 65$.

A square of order 5 in which any pair of numbers opposite the center adds up to $n^2 + 1$ (n

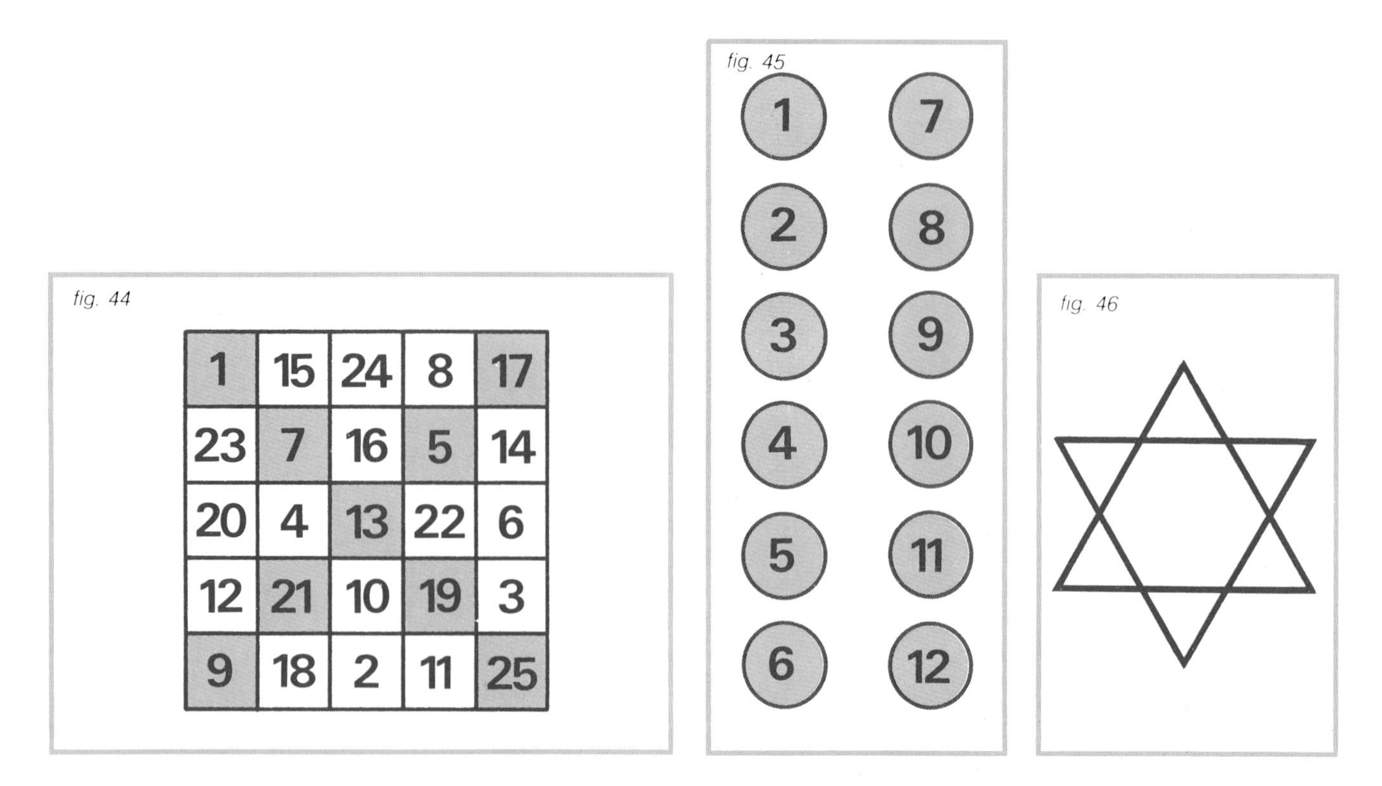

fig. 44

fig. 45

fig. 46

being the order) is called "associative." Here, $n^2+1=26$. Thus, horizontally, 20 is opposed to 6, and $20+6=26$; diagonally, 17 is opposed to 9 and $17+9=26$, 25 is opposed to 1, and $25+1=26$. The Lo shu of Fig. 29 also has this property and is therefore associative. Indeed, with $n=3$, $n^2+1=10$, and in the square $7+3=10$, $4+6=10$, $8+2=10$, $9+1=10$. A fourth-order square may be either associative or pandiagonal, but never both. The smallest square that can be both is of the fifth order. If, as usual, we exclude rotations and reflections, there are only 16 fifth-order squares with both properties according to Schroeppel's calculations.

In Medieval times, the Moslems imbued pandiagonal squares of order 5 with 1 at the center with mystic significance, for number 1 is the sacred symbol of Allah, the Supreme Being. The problem of representing God and the concept of God occurs in all religions and theologies. The symbol that best evokes the unity of being is the number 1. God is one. However, the Moslem conception of God is such that no sign or picture can adequately represent Him, not even the most abstract and immaterial such as the number 1. Hence, in some magic squares the ineffable nature of the Supreme Being is suggested by leaving the central square empty.

Magic stars

Similar features are observed in other geometrical figures such as magic stars. Take twelve counters numbered from 1 to 12 (Fig. 45) and construct a star of David from two equilateral triangles (Fig. 46). Now place the counters on vertices and intersections so the numbers along each of the six sides add up to the same sum. As before this can be achieved by trial and error. In Fig. 47, the sum is always 26. However, if we add up the six vertices we get $3+2+9+11+4+1=30$.

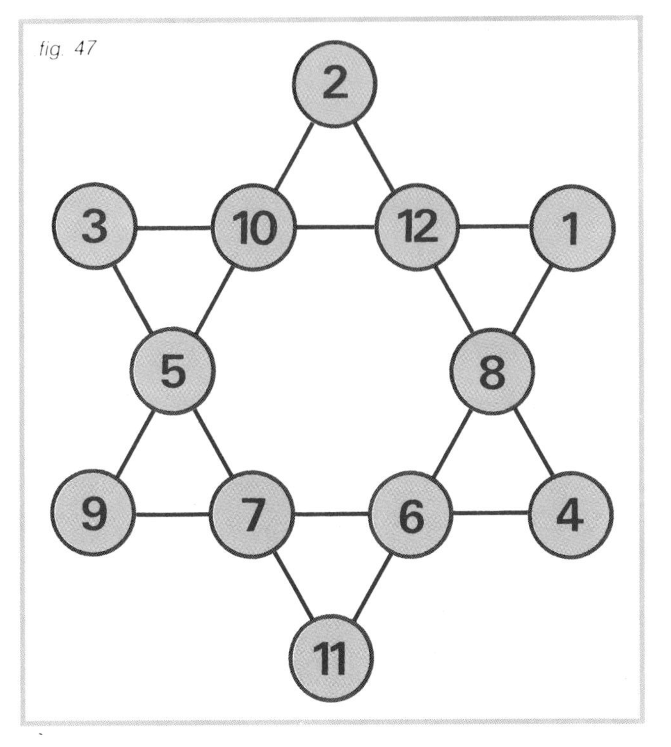

fig. 47

Below: A Chinese magic circle of great historical interest, executed by Seki Kowa in the 17th century.

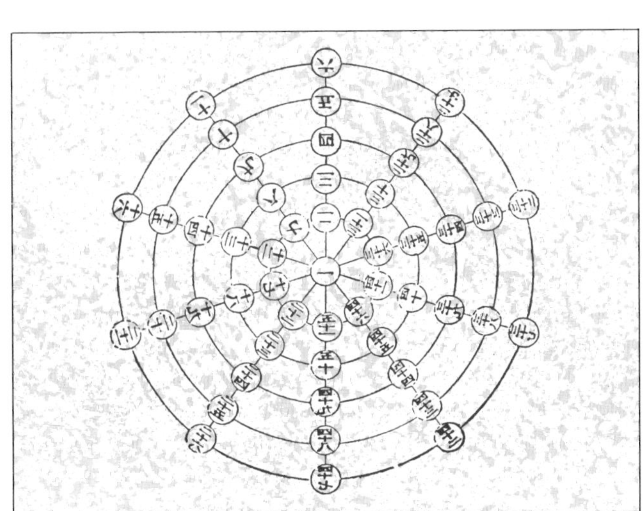

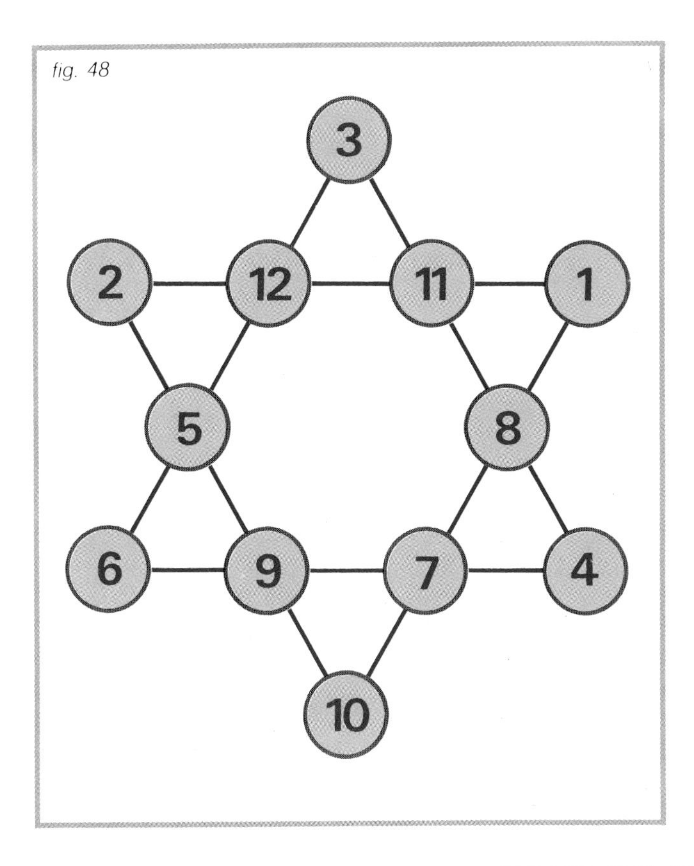

fig. 48

Let us refine the star by displacements that make this last sum also equal to 26. Such a problem must be tackled systematically and demands a plausible strategy. With the figure consisting of two equilateral triangles, in order to reach a sum of 26 at the vertices, we must put the internal numbers aside for the moment (they form a hexagon shared by the two triangles). A rational procedure might be to produce the vertex sum of 13 for each triangle, so that $2 \times 13 = 26$. In Fig. 47, the inverted triangle gives a vertex sum of $11 + 1 + 3 = 15$. We therefore interchange 11 and 10, 3 and 2. The vertex sum becomes $10 + 1 + 2 = 13$. For the upright triangle we can no longer use 1 or 2 (indeed $9 + 4 + 2 = 15$), or 8 or 7; they would merely complicate matters. Therefore we try 6; we interchange 6 and 9, 3 and 2, and leave 4 untouched. A small rearrangement on the sides then produces the solution shown in Fig. 48.

fig. 49

19	33	20	18	17
13	27	14	12	11
6	20	7	5	4
15	29	16	14	13
9	23	10	8	7

fig. 52

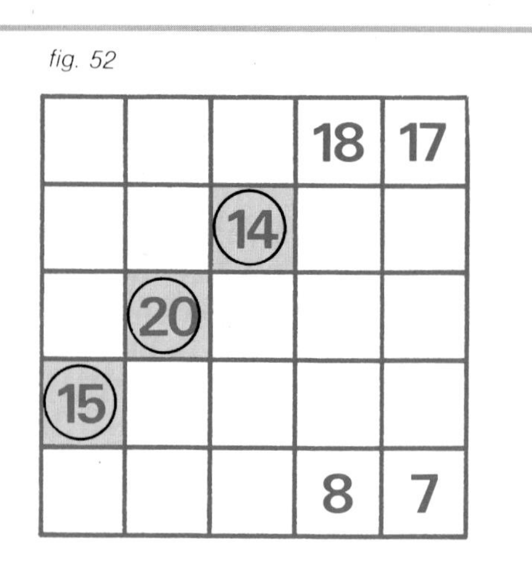

			18	17
		14		
	20			
15				
			8	7

fig. 50

19		20	18	17
13		14	12	11
	20			
15		16	14	13
9		10	8	7

fig. 53

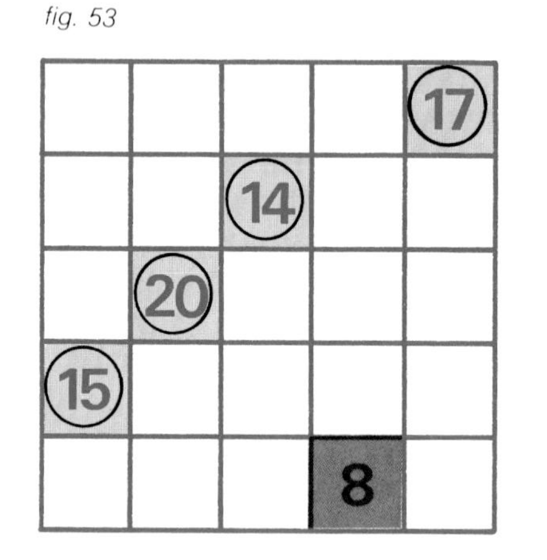

				17
		14		
	20			
15				
			8	

fig. 51

19			18	17
		14		
	20			
15			14	13
9			8	7

	2	16	3	1	0
17	19	33	20	18	17
11	13	27	14	12	11
4	6	20	7	5	4
13	15	29	16	14	13
7	9	23	10	8	7

fig. 54

The geometric representation of the principles of yin and yang (female and male, according to ancient Chinese philosophy; see p. 42). Ancient peoples tended to represent ultimate principles through abstract, stylized or geometrical figures.

More about squares

There are still more magic squares. Take Fig. 49. Its square seems to have no formation rule, its numbers being haphazardly distributed. However, the square has a property that furnishes some interesting tricks. Ask a player to perform the following:

1) Take any number and eliminate all others in the same row and column. For this we need counters or other markers to cover the numbers to be removed. Suppose the number chosen is 20 in the third row and second column. Eliminate 6, 7, 5, 4 and 33, 27, 29, 23, leaving the square of Fig. 50.

2) Repeat the maneuver on Fig. 50, and suppose 14 is chosen. Eliminate 13, 12, 11 and 20, 16, 10, leaving Fig. 51.

3) Repeat as before and suppose 15 is chosen. Eliminate 14, 13 and 19, 9, leaving Fig. 52.

4) Repeat and suppose 17 is chosen. Eliminate 18 and 7, leaving Fig. 53. Only 8 remains. Adding this and the four chosen numbers $(20 + 14 + 15 + 17 + 8)$ we get 74. Repeating the whole procedure with any other numbers, the result will always be 74. What is the trick?

Consider how the square is constructed. Any number is the sum of two, one each from a group of generators that together add up to 74 (Fig. 54): $2 + 16 + 3 + 1 + 0 + 17 + 11 + 4 + 13 + 7 = 74$. The two groups are shown in black along the first row and column; any number of the square is the sum of the generators against its row and column.

The trick then is to eliminate all numbers except one (and only one) in each row and column, and that is achieved by the procedure stated above. The final sum then, is simply the sum of the two groups of generators; a rather simple device in which the order of the square does not matter, nor does the sum to be calculated. Any type of numbers can be used: negative, positive, fractions or integers.

An extraordinary surface

There are mathematical and geometric games that can be resolved only by a proof or through a concrete example. Take a square with 16-inch sides for instance, and subdivide it into four as in Fig. 55. We can then transform it into the rectangle of Fig. 56. The four parts fit perfectly, yet the two figures are unequal in area, for $16 \times 16 = 256$ in^2 and $10 \times 26 = 260$ in^2. It appears that we magically produced 4 in^2 out of nothing. Here, too, there is a trick, as we can see by actually constructing the figures. Take a large sheet of graph paper, large-meshed if possible, and substitute for each inch a certain number of squares (Fig. 57). Suppose the square has 8-inch sides, so there are $8 \times 8 = 64$ small squares. Cut out the pieces as required (Fig. 58). The rectangle will be $5 \times 13 = 65$, leaving us with one too many squares. If we lay out the pieces as in Fig. 59, it is at once clear that the sides of the two triangles and trapezia do not form a diagonal of the rectangle: it is only the diagram that produces this illu-

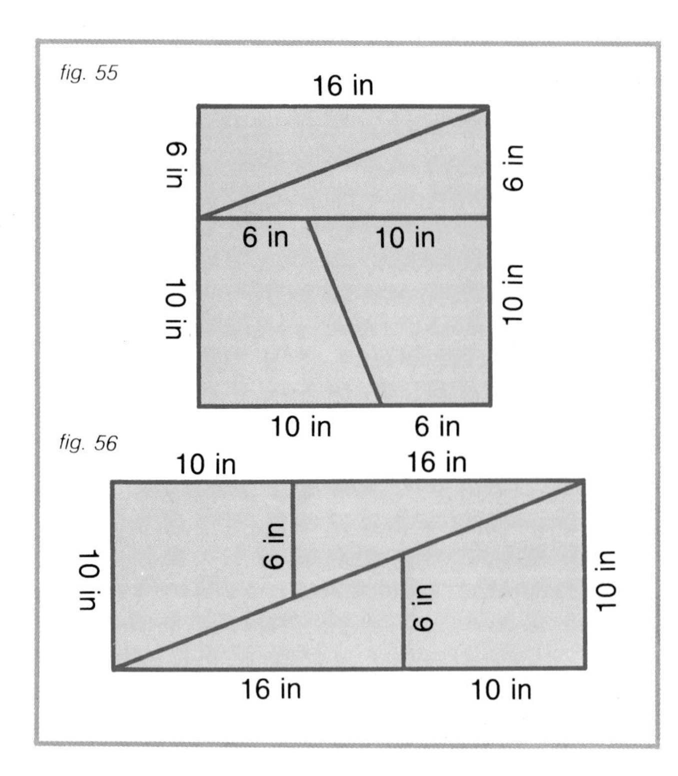

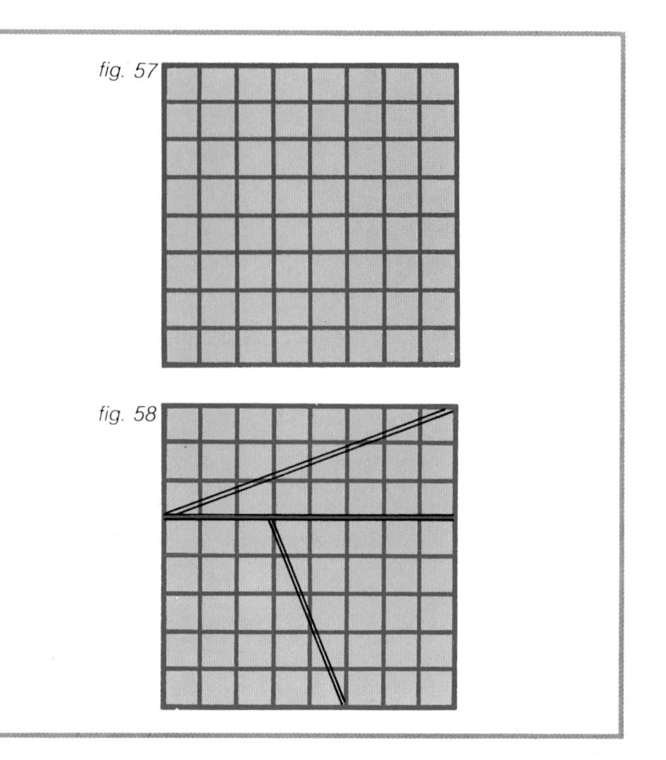

sion. Actually there is a gap of an internal parallelogram of area 1.

The bridges of Königsberg

This large East Prussian town (now called Kaliningrad and part of the Soviet Union) lies on the river Pregel which, in the 18th century, was crossed by seven bridges linking the various sections, as shown in Fig. 60. The town is best known as the birthplace of Immanuel Kant (1724–1804), the noted German philosopher. However, mathematicians know Königsberg because its layout is the basis for an intricate puzzle which, in Kant's time, eluded even the most famous of them. The problem is this: Like the inhabitants of other German cities, the Königsbergers strolled through town on Sundays; was it possible to plan such a walk so that setting out from one's house one could return to it after crossing each bridge once and only once? The Swiss mathematician Leonhard Euler (1707–1783), born in Basle, studied the problem and finally answered, no! (His research, originally involving puzzles of this kind, laid the foundation for a new branch of mathematics, the theory of graphs. An elementary account of the theory is given in the next section.) Here is Euler's general rule to determine the solution to this and similar problems. We count how many bridges end on each bank or island. If more than two totals are odd, there is no solution. If the totals are even—or only two are odd—there is a solution, namely a path crossing each bridge once and only once. Consider Fig. 60. On A, there are three bridgeheads; on B, 3; on C, 5; on D, 3. More than two totals are odd, hence there is no solution.

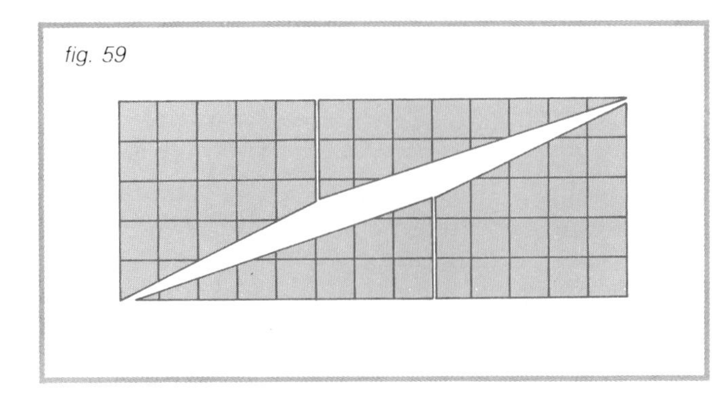
fig. 59

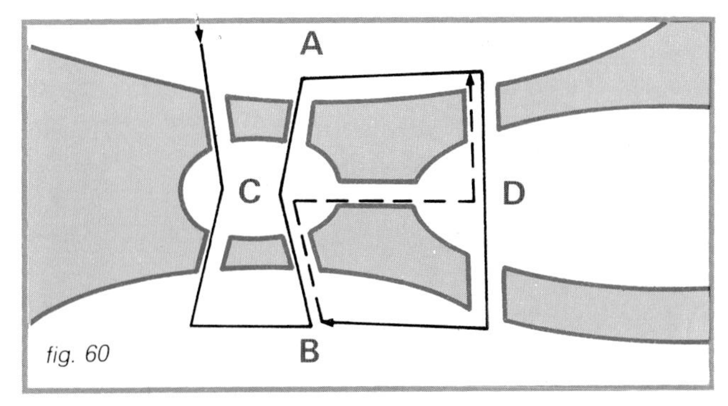

fig. 60

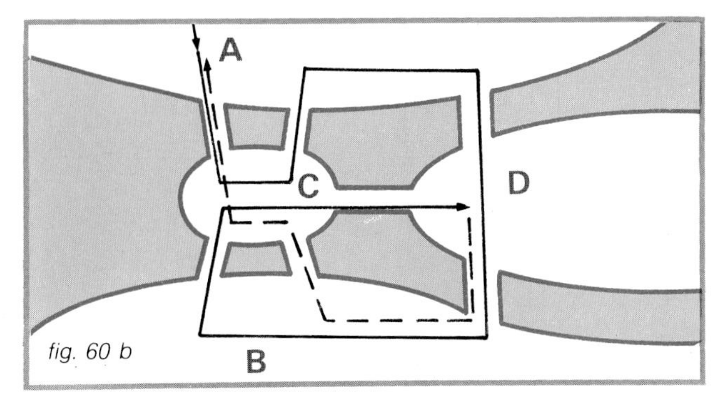

fig. 60 b

Below: A 17th-century engraving showing Königsberg, with the river Pregel and its seven bridges. It was such maps that generated the study of topology. Can we traverse the city's seven bridges in such a way that we cross each one only once? In 1735 the Swiss mathematician Euler used his theory of graphs to prove that it cannot be done

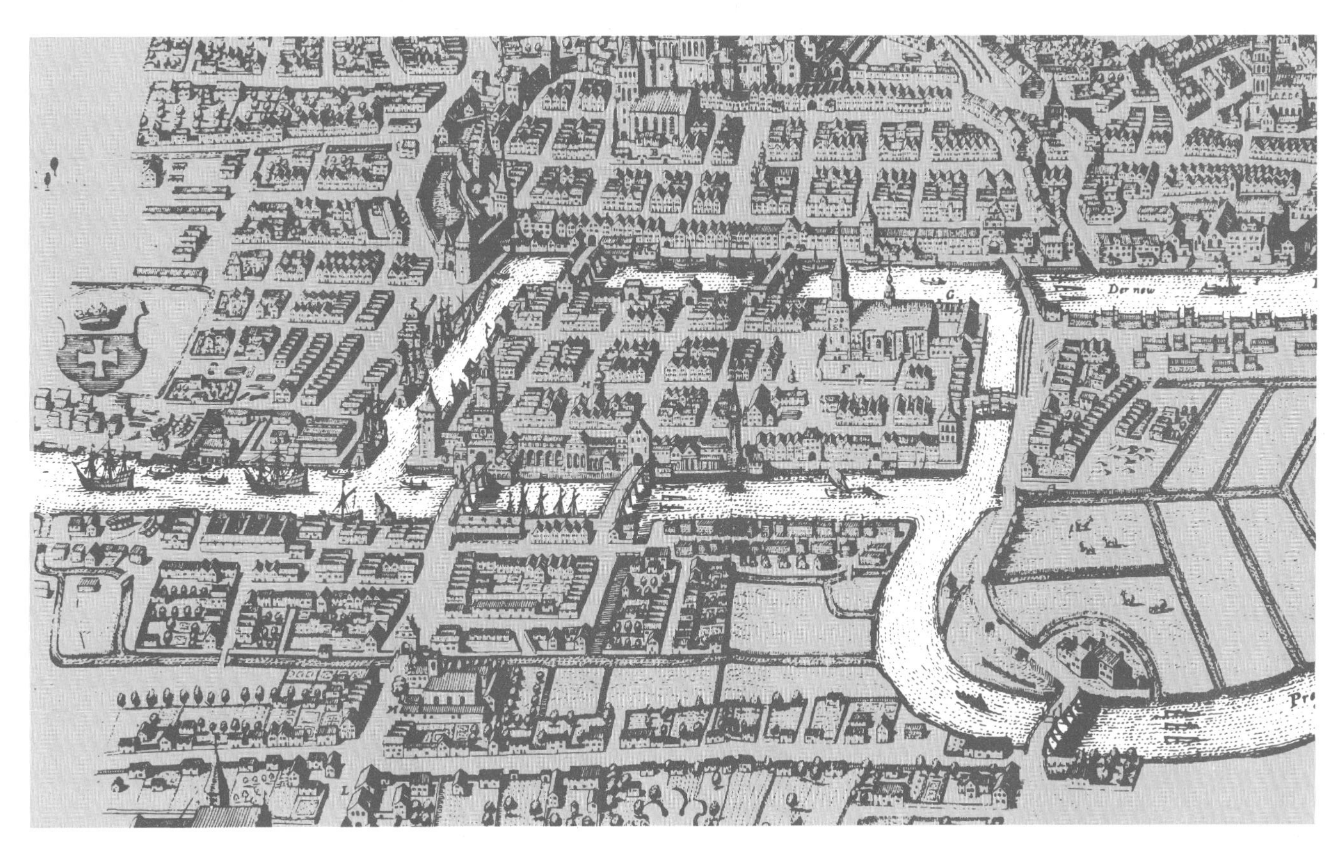

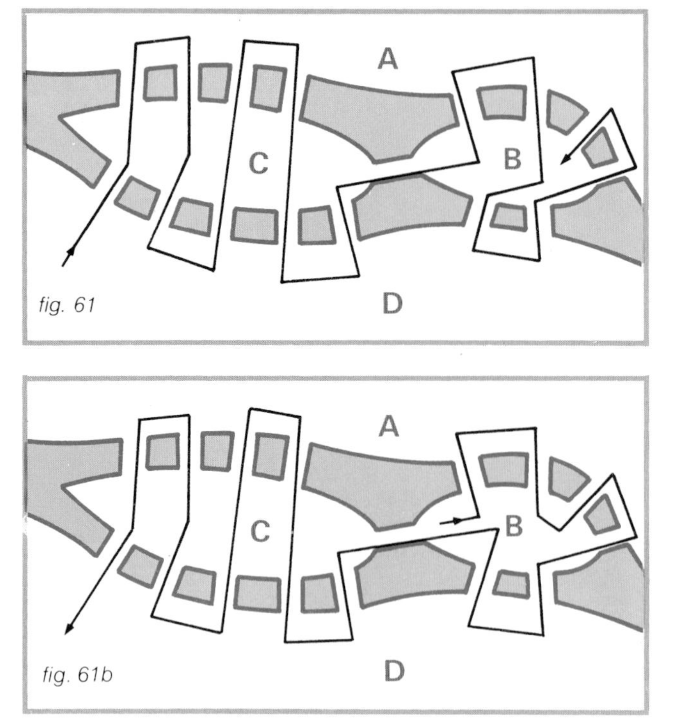

fig. 61

fig. 61b

Euler's work on graphs appeared in 1736. Since then it has been usefully applied not only in mathematics but in other fields as well. In the 19th century, graphs were used in circuitry and in theories of molecular diagrams. Today, aside from being a method of analysis in pure mathematics, the theory of graphs is used for the solution of numerous practical problems, for example in transportation and programming.

Euler was one of the most productive and original mathematicians in the history of science. The son of a Calvinist pastor, he was barely twenty when, in 1727, he was invited to join the Academy of Science in St. Petersburg (today's Leningrad). He had an encyclopedic mind and though a student of physics, astronomy and medicine, Euler had a particular fondness for mathematical problems. His output was prodigious. It is said that he wrote constantly—while waiting for dinner to be served, even while holding one of his many offspring. Indeed his desk was always laden with work awaiting publication. In 1746 Euler lectured at the Berlin Academy, but finding the cultural climate and the appreciation of his work less than

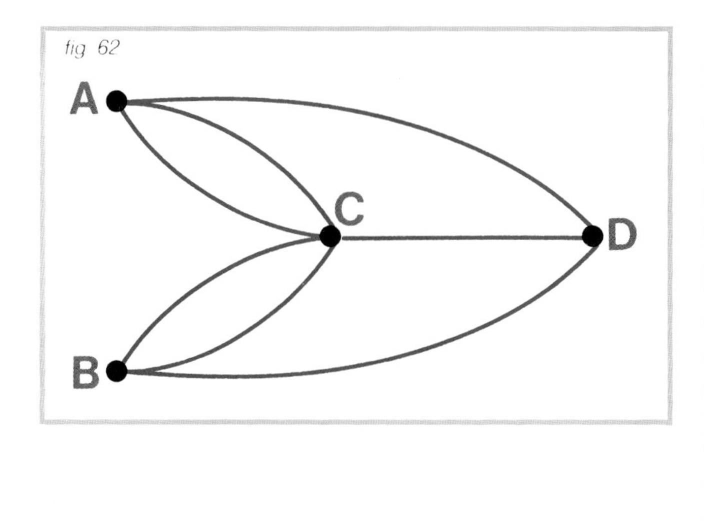

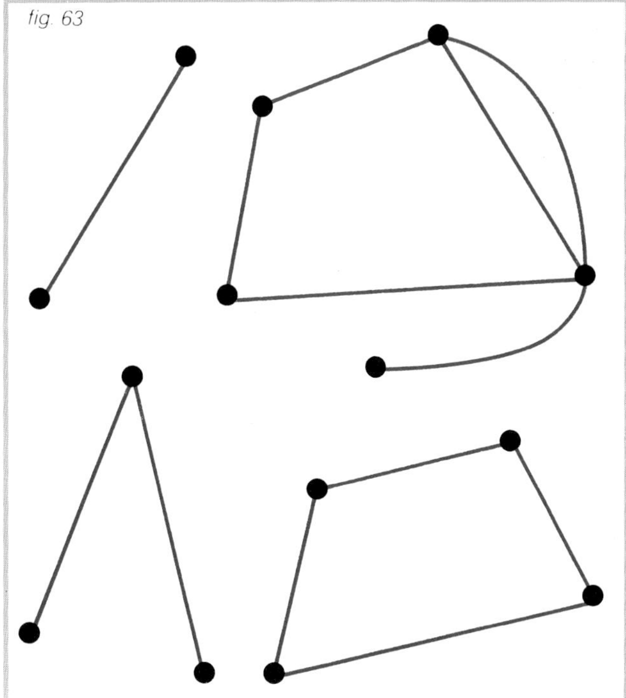

Opposite right: A 17th-century print of Paris showing the Seine and the Ile de la Cité. A problem similar to that of Königsberg can be posed, but, in this instance, the theory of graphs proves that the problem of the bridges can be solved.

favourable, he returned to Russia and the court of Catherine the Great. Even though Euler eventually went blind, he pursued his mathematical researches intensely until shortly before his death in 1783. Some time ago, Swiss mathematicians honoured Euler by beginning to collect and publish all his writings; some fifty volumes have appeared to date, and the total may well reach two hundred.

The bridges of Paris present a problem similar to the bridges of Königsberg. Consider the Ile de la Cité in the Seine (Fig. 61). Here *A* has 8 bridgeheads; *B, 7; C*, 10; *D*, 7; producing only two odd totals. Therefore, there must be a solution, but with certain restrictions as we shall see. The solution is easily found by starting from an area with an odd total and tracing a path crossing the greatest number of pairs of bridges leading from one area to another. A further solution is shown in Fig. 61b.

Elementary theory of graphs

When Euler grappled with the problem of Königsberg's bridges, he did not consider going there to solve it. Instead, in the manner of modern science, he tried to formulate the problem in a general manner by tracing a schema (Fig. 62) in which banks and islands are shown as points, and the various bridges between them as lines. The problem then is: Starting from any of the points, trace the figure and return to the same point without retracing any line and without lifting pen from paper. It is impossible. To solve the problem formally, consider some auxiliary concepts first. What does a graph amount to? Given two or more points in a plane, we join them with arcs or curves or segments to obtain a figure we call a graph. The points are called vertices or nodes, and the lines between them (of whatever shape) are called sides or edges. The number of

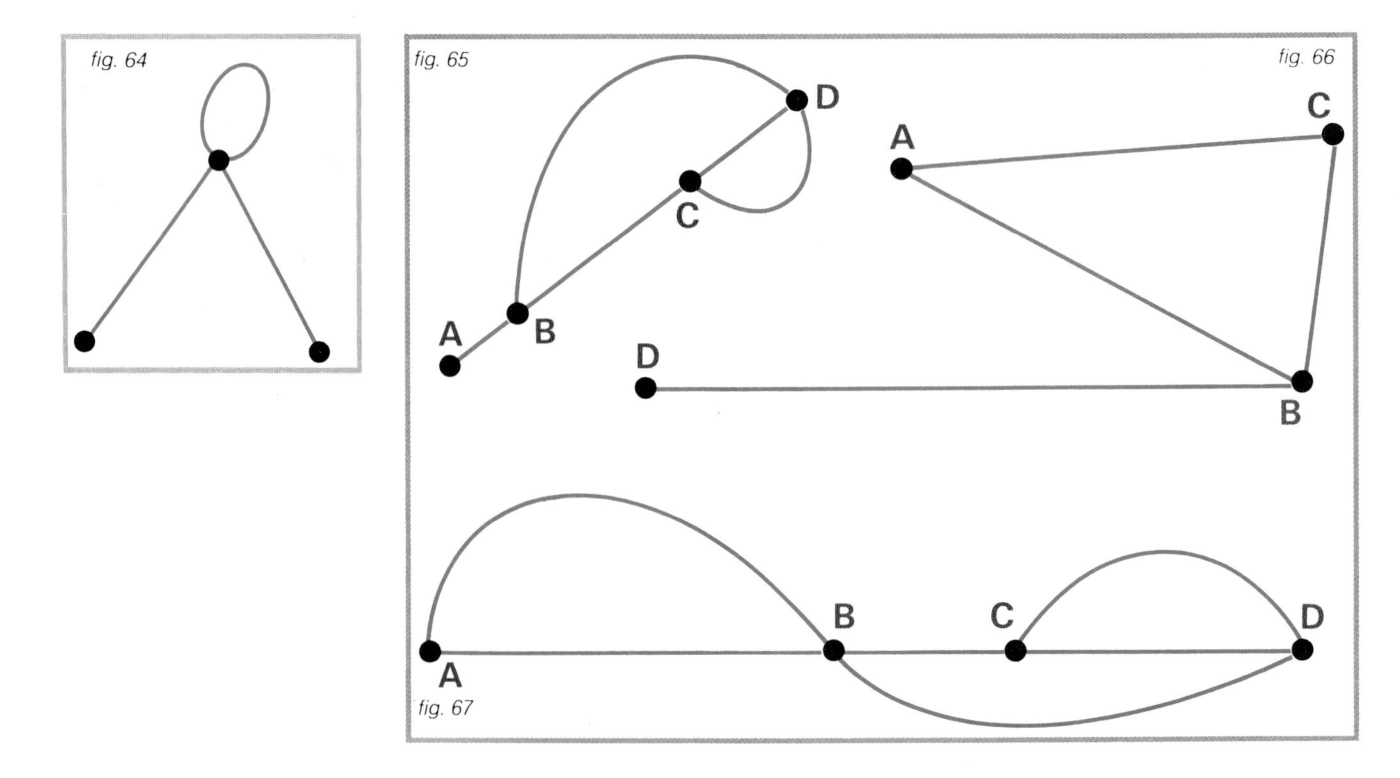

vertices is known as the order of the graph. Fig. 63 gives some examples. The term "graph," will be used here with the particular meaning defined above, and not in its more general sense. In some graphs, the initial and final vertex of an edge can be the same. Sides that link a vertex with itself are referred to as loops (top of Fig. 64).

What is important in a graph is its order, namely the number of vertices, and the links between them. In other words, graphs are not differentiated by their shape. Those in Figs. 65–67, for example, are equivalent, or isomorphic (from the Greek word meaning "same form") because they have the same number of vertices and the same links. Another basic concept is that of the order of a vertex, which is defined as the number of links that end at that vertex. Accordingly, there are odd and even vertices. In Fig. 67, *C* and *D* are of order 3, while in Fig. 66, *A* is of order 2 and *B* is of order 3. Thus a graph may have only even vertices, or only odd ones, or some of each, although in the last instance the number of odd vertices must be even. Try and draw a graph with an odd number of odd vertices!

We need one further concept. A graph is called traversable (in the sense of an Euler path or line) when each side is traversed only once; vertices, on the other hand, may be traversed any number of times. Hence the following rules discovered by Euler:

1) If a graph has only even vertices, it is traversable from any one vertex back to that vertex.

2) If a graph has only two odd vertices, it is traversable but without returning to the initial vertex.

3) If a graph has more than two odd vertices (four, six, eight and so on) it is not traversable.

Returning to the problem of Königsberg and its bridges, there are four odd vertices, therefore the problem cannot be solved.

A closed trace in which each vertex is passed only once is called a Hamilton circuit, after the Irish mathematician William Rowan Hamilton (1805–1865) who first showed that such circuits exist. For example, the graph consisting of the

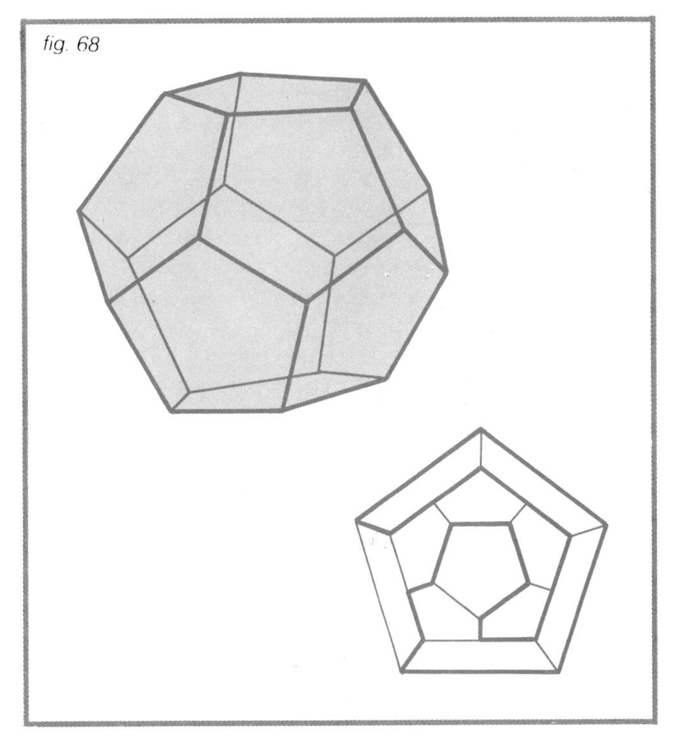
fig. 68

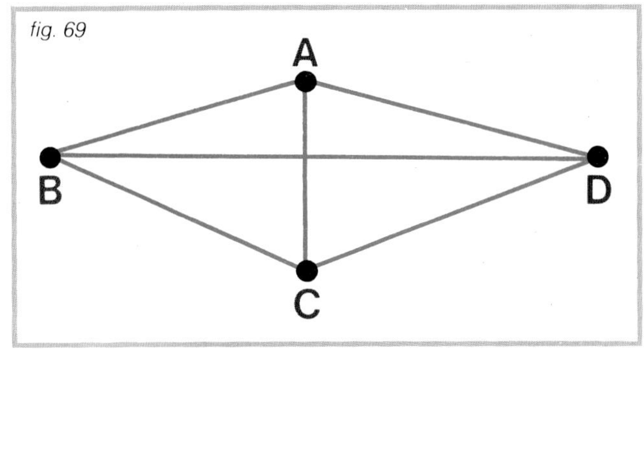

fig. 69

vertices and edges of the regular dodecahedron has a Hamilton circuit (Fig. 68). A classic example is the following: Let the vertices *A*, *B*, *C*, *D* stand for four cities (Fig. 69). What are the possible paths that go through each vertex once and once only? Starting from *A*, we have the following Hamilton routes: *ABCDA*, *ABDCA*, *ACBDA*, *ACDBA*, *ADBCA*, *ADCBA*. Note that 1 and 6, 2 and 4, 3 and 5, are pairs differing only in direction.

Save the goat and the cabbage

This is an old saying, but not everyone knows that its origins are an ancient puzzle of some twelve centuries ago. A man wants to transport a wolf, a goat, and a cabbage across a river in a boat that barely has room for him and the cabbage, and certainly for not more than one of the animals. Moreover, he cannot leave the wolf alone with the goat, or the goat with the cabbage. How can he get everything across the river without the wolf eating the goat, or the goat devouring the cabbage?

Graphs are essential for solving those puzzles in which we must move from one place to another under certain conditions. Let us represent the various crossings, denoting the man by *t*, the wolf by *l*, the goat by *p*, and the cabbage by *c*. The first trip might be to take the goat across, since the wolf will not eat the cabbage. Starting with the group *tlpc*, we now have *lc* left. Next, *t* returns alone which creates the group *tlc*. Now *t* transports either the wolf or the cabbage. In either case, he returns with the goat, so the group is now *tpc* or *tlp* respectively. Now he takes the cabbage if he had already taken the wolf, or the wolf if he had already taken the cabbage. When the man returns alone to join *p*, the group becomes *tp*. They finally cross the river and that concludes the operation.

A synthetic graph for these moves is shown in Fig. 70. This simple example enables us to visualize a graph as a game. The vertices represent the various positions (the changes in the original

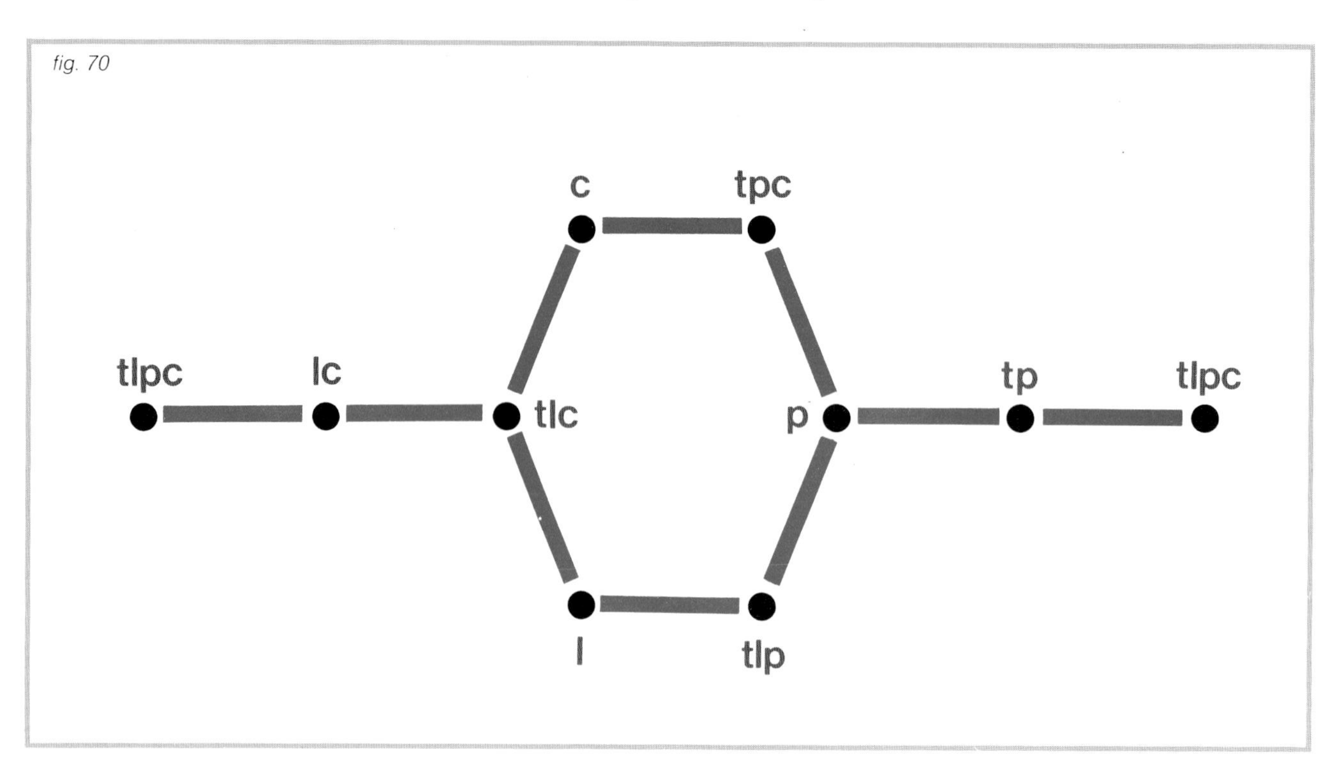

group), and the lines between them are the moves allowed.

To clarify the advantages of graphs still further, try to solve the present problem intuitively by drawings (Fig. 71). This turns out to be more complicated, so for the next game we will only use graphs.

The jealous husbands

This is similar to the last problem and just as old but a bit more complex. Three honeymoon couples reach a river and find a small boat that will hold only two people. The dilemma is made worse by the fact that the husbands are rather jealous. How can the entire party cross the river without leaving any bride alone with a man who is not her husband? As before, let us simplify the problem and construct a graph. Let the couples be A, B, C, and men be distinguished from women by the suffixes, u and d respectively. Thus a_u and a_d represent husband and wife of couple A, so that using a synthetic notation $A = (a_u, a_d)$, $B = (b_u, b_d)$, $C = (c_u, c_d)$.

The first vertex will be given by A, B, C, or all three couples together. The problem is knottier than the previous one because there are more combinative alternatives, as can be seen in detail in Fig. 72. First, two wives cross the river, yielding three possible groups—a_d, b_d or b_d, c_d, or c_d, a_d. From the first vertex there are three sides to show this. At successive vertices we show the changes of the original group until everyone has crossed. Thus, A, b_u, C means that on the starting bank we now have everybody except the wife of couple B. The vertex marked a_u, b_u, c_u means that on the starting bank we have all the men, while all the women are on the opposite bank. The fact that all the other sides converge here indicates the ne-

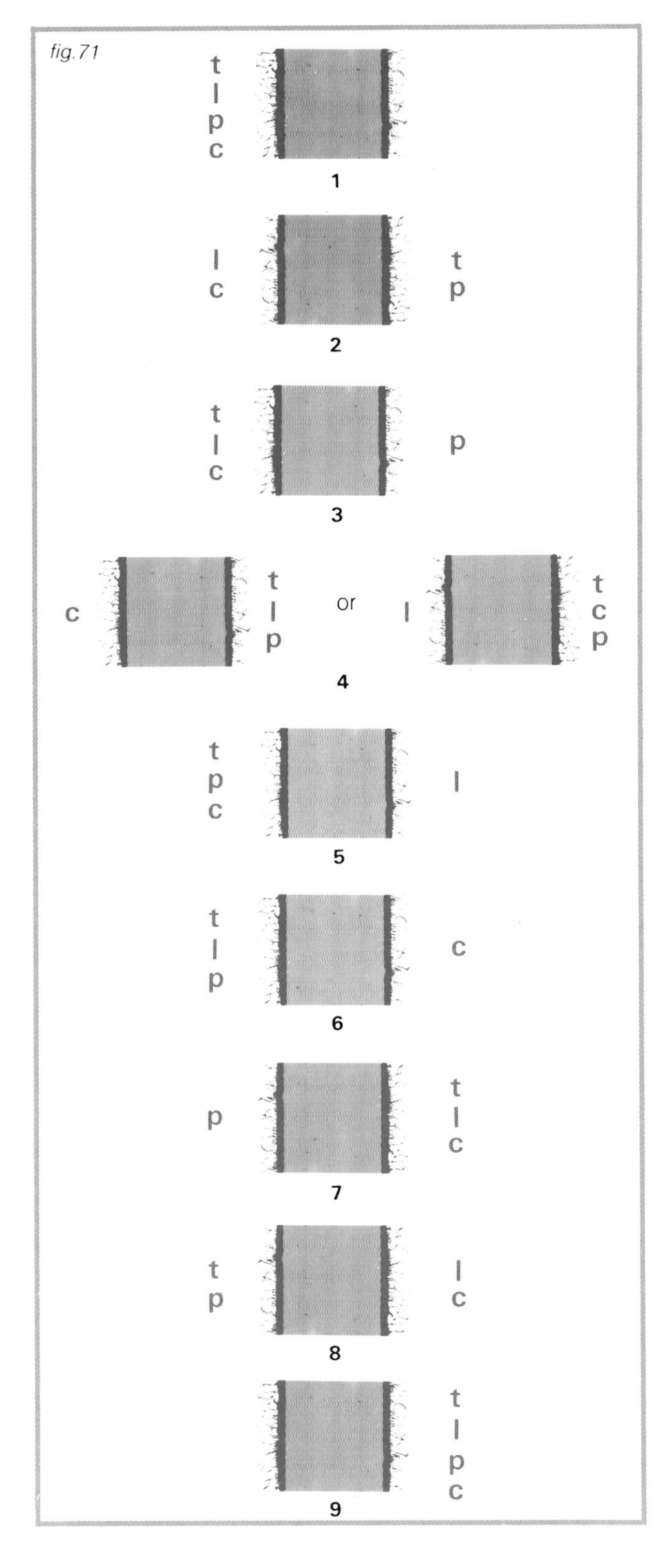

Left: 1) Starting position with everyone on the left bank; 2) the man takes the goat to the right bank; 3) he returns alone, leaving the goat; 4) he takes either the wolf or the cabbage to the right bank; 5) if the former, he returns with the goat and leaves the wolf; 6) if the latter, he returns with the goat and leaves the cabbage; 7) he leaves the goat on the left and ferries either the cabbage or the wolf, (depending on whether 5) or 6) was the case), to the right bank; 8) he returns to the left bank to fetch the goat; 9) final position with everyone on the right bank.

cessity of this stage. Indeed, one of the two women who had crossed first returns (producing A, b_u, C or A, B, c_u) and helps the remaining woman to embark, leaving the three men (a_u, b_u, c_u) alone. One woman then disembarks while the other returns to the three men. We now have a complete couple (either A, or B, or C) along with the two other men. The next move is: The two men embark, leaving the couple. One man disembarks on the other bank while the second man returns with his own wife, producing two couples (A, B, or B, C, or C, A) at the starting point. The two men embark leaving their wives. From the other bank, the third woman, who had been there with her husband, returns alone, and in two further crossings brings the other two women across, thus reuniting the entire party.

Interestingly, even using identical rules, when the same problem involves four instead of three couples it is unsolvable. Remember that only the men are jealous, which means that on neither

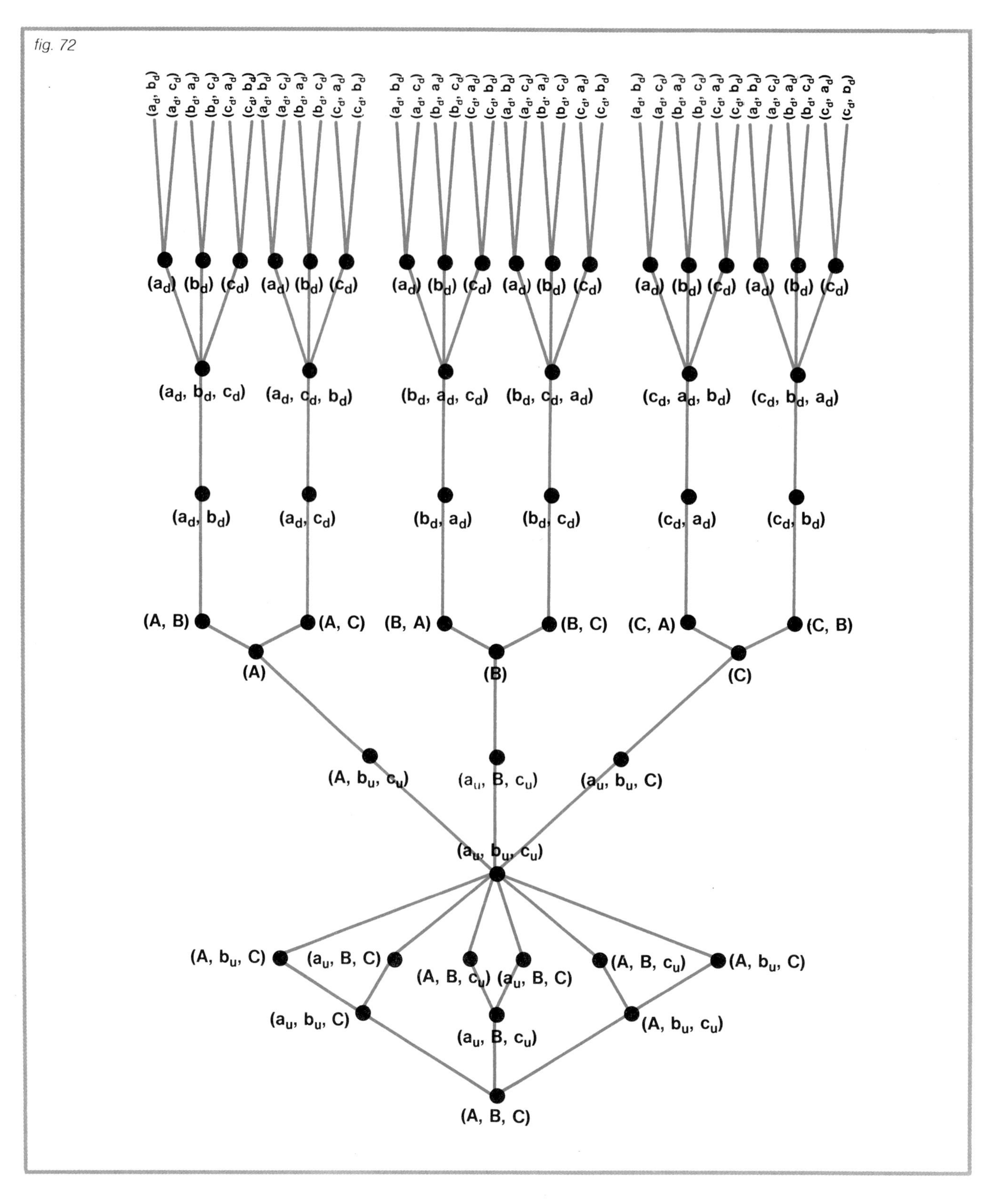
fig. 72
(A, B, C)
(A, b_u, C)
(a_u, B, C)
(A, B, c_u)
(a_u, B, C)
(A, B, c_u)
(A, b_u, C)
(a_u, b_u, C)
(a_u, B, c_u)
(A, b_u, c_u)
(a_u, b_u, c_u)
(A, b_u, c_u)
(a_u, B, c_u)
(a_u, b_u, C)
(A)
(B)
(C)
(A, B)
(A, C)
(B, A)
(B, C)
(C, A)
(C, B)
(a_d, b_d)
(a_d, c_d)
(b_d, a_d)
(b_d, c_d)
(c_d, a_d)
(c_d, b_d)
(a_d, b_d, c_d)
(a_d, c_d, b_d)
(b_d, a_d, c_d)
(b_d, c_d, a_d)
(c_d, a_d, b_d)
(c_d, b_d, a_d)
(a_d) (b_d) (c_d)

fig. 73

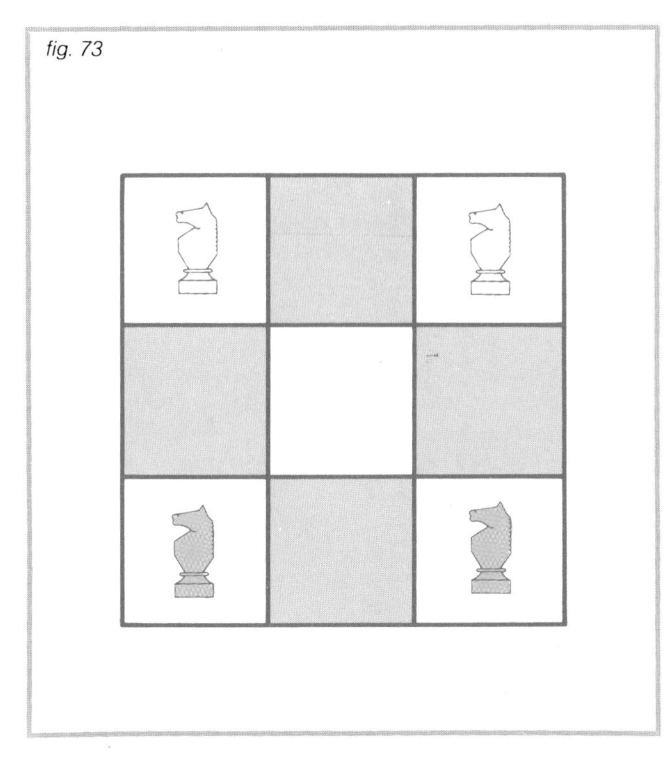

fig. 74

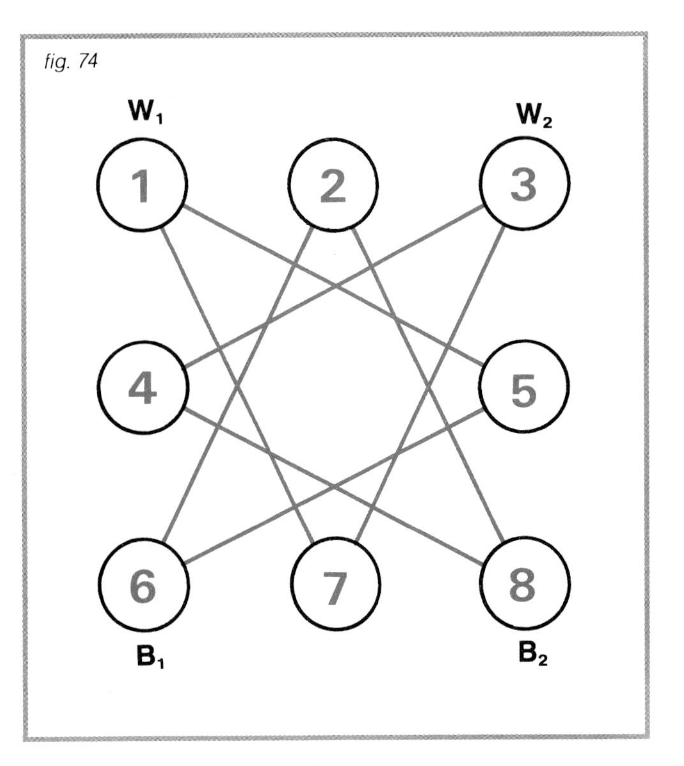

bank can women be left with husbands who are not their own, even if these husbands have their wives with them.

To review the individual alternatives and to show that each is unsolvable would take too long. Instead, let us use the following shortcut. At each crossing the number of people transported will increase. Upon reaching a certain point there must be five people. There could be:

1) 4 women and 1 man,

2) 3 women and 2 men,

3) 2 women and 3 men, or

4) 1 woman and 4 men.

1) and 2) are ruled out, for at least one of the women would be without her husband. Nor will 3) do, for that would leave 2 women and 1 man on the other bank and hence one woman without her husband. There remains 4), but to attain a group of 1 woman and 4 men, those who had just arrived must have been either 1 man and 1 woman, or 2 men. If the first is so, there must have been 1 man and 4 women on the starting bank, which leads back to 1); if the latter is so, the 2 men must have left 3 women, which leads back to 2). Thus the puzzle is unsolvable.

A simple change in the initial conditions, however, makes the problem possible even with four couples. If we have a boat holding three people instead of two, then one woman can take the other women across and return to meet her husband while the other men join their respective wives. Now the solution is easy and we leave the reader to work out the remaining moves.

Interchanging knights

This problem was invented and solved by the British mathematician Henry Ernest Dudeney (1857–1931) some sixty years ago. A simple graph lets us visualize the problem and reach the solution at once. Trial and error would lead to serious difficulties.

Draw the reduced chessboard of Fig. 73. The game consists of interchanging the two white knights with the two black ones. Given the

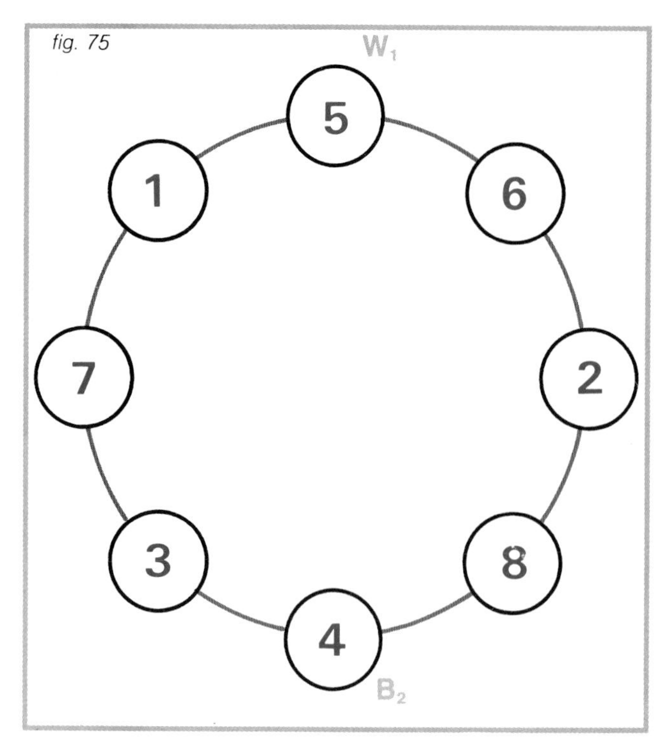

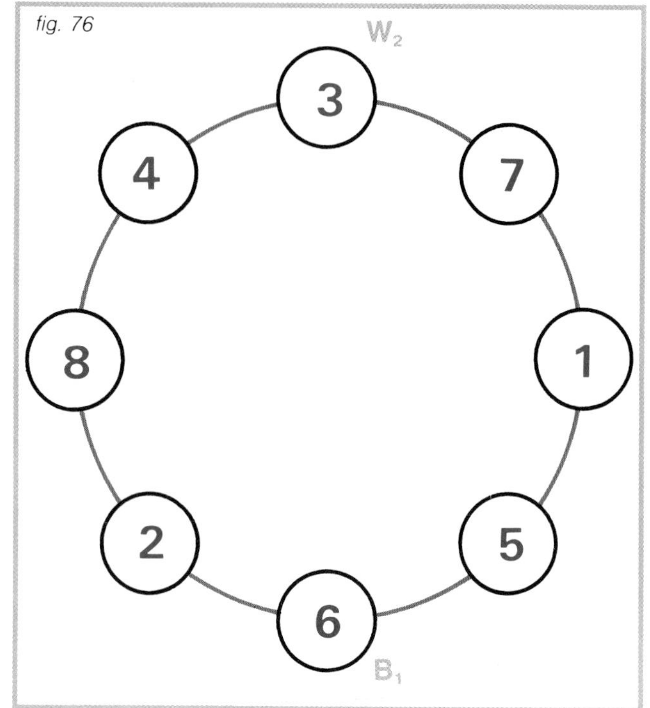

knight's move, we can construct the graph of Fig. 74, where W_1 and W_2 are the two white knights, and B_1 and B_2 the two black ones. The central square is not numbered because it is inaccessible to the knights. The graph of Fig. 74 is clear enough. For example, to shift W_2 to square 6, the piece must follow the route 3-4-8-2-6, while B_1 reaches 3 by 6-5-1-7-3. Further, to shift W_1 to 8, the route is 1-7-3-4-8, while B_2 traverses 8-2-6-5-1.

In the graph of Fig. 74 the sides intersect at various points that should not be considered vertices. If a graph can be drawn without such intersections it is called planar. The following is a graphic solution of the knight interchange by planar graphs. The only restriction is that we must isolate the moves of symmetrically opposite knights (Figs. 75–76). The graphs can be read in either direction, clockwise or counterclockwise.

A wide range of applications

The theory of graphs, born from giving mathematical forms to puzzles and first used in geometry and mathematics, was applied in numerous areas of science as it was in practical life. Because of its formal properties, it was swiftly developed as a way to simplify and visually present otherwise complicated problems. Since the last century, it has proven immensely fruitful in representing problems of electric circuitry.

Fig. 77 shows the graph for combining three switches with two lamps. The arrows indicate that the sides are oriented; a graph is directional when its sides can be traversed only by following arrows. If the sides are not directional we can run along them in either direction and the figure will have no arrows. Graphs serve a number of purposes. Among other things, they are widely used for road routing, floor plans and economic programming.

Because graphs are simple and immediate, they are often used to visualize such complex situations as relations between people or groups of people. The stages of a football championship in which a number of teams play each other might be an example. Another might be a simple routing problem: the road between London and Dover. In

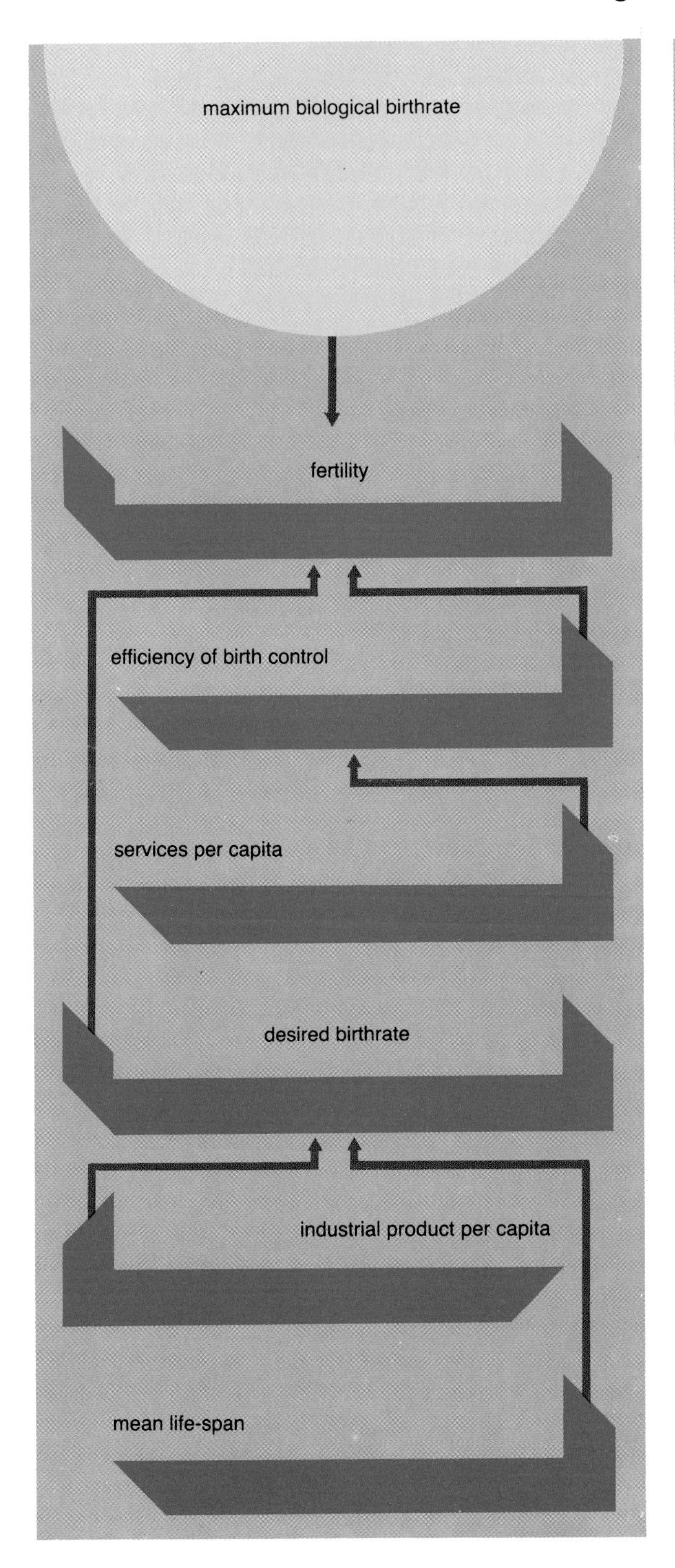

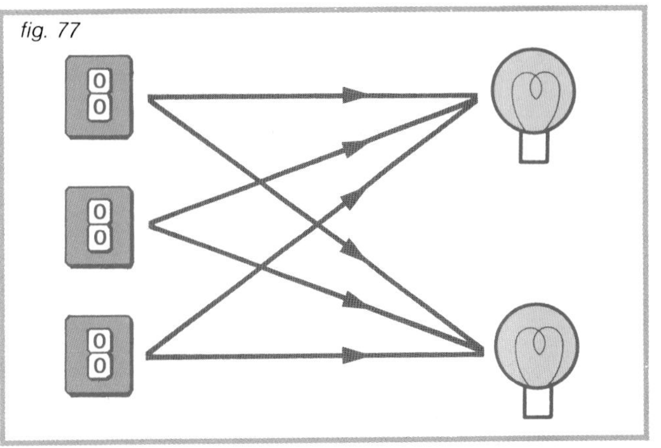

Graphs simplify and clarify situations and have become essential tools in many fields.
Left: A graphic description of the factors affecting the world's birthrate. Fertility depends on two factors: the efficiency of birth control, and the desired birthrate. These in turn depend on other factors: per capita services and industrial product, and the average life-span.

this case, two vertices would represent the two cities, and the sides would indicate any two roads between them. If we take part of a superhighway, its separate lanes are directional and marked with one-way arrows. Of course many ordinary roads are also two-way, and in that sense can be driven on only one half of the road, but on a superhighway the division between the two directions is more apparent. In some cases, the route may return to the starting point without passing other vertices, and then it is called a loop. Graphs are the best method for tackling the complex problems of routing. Finally, consider the problem of representing traffic conditions in a large city and its various individual areas with their two-way and one-way systems and "No Turns," "Do Not Enter," and "Detour" signs.

Topology, or the geometry of distortion

The theory of graphs presented briefly here is only one branch of topology, a recent and partic-

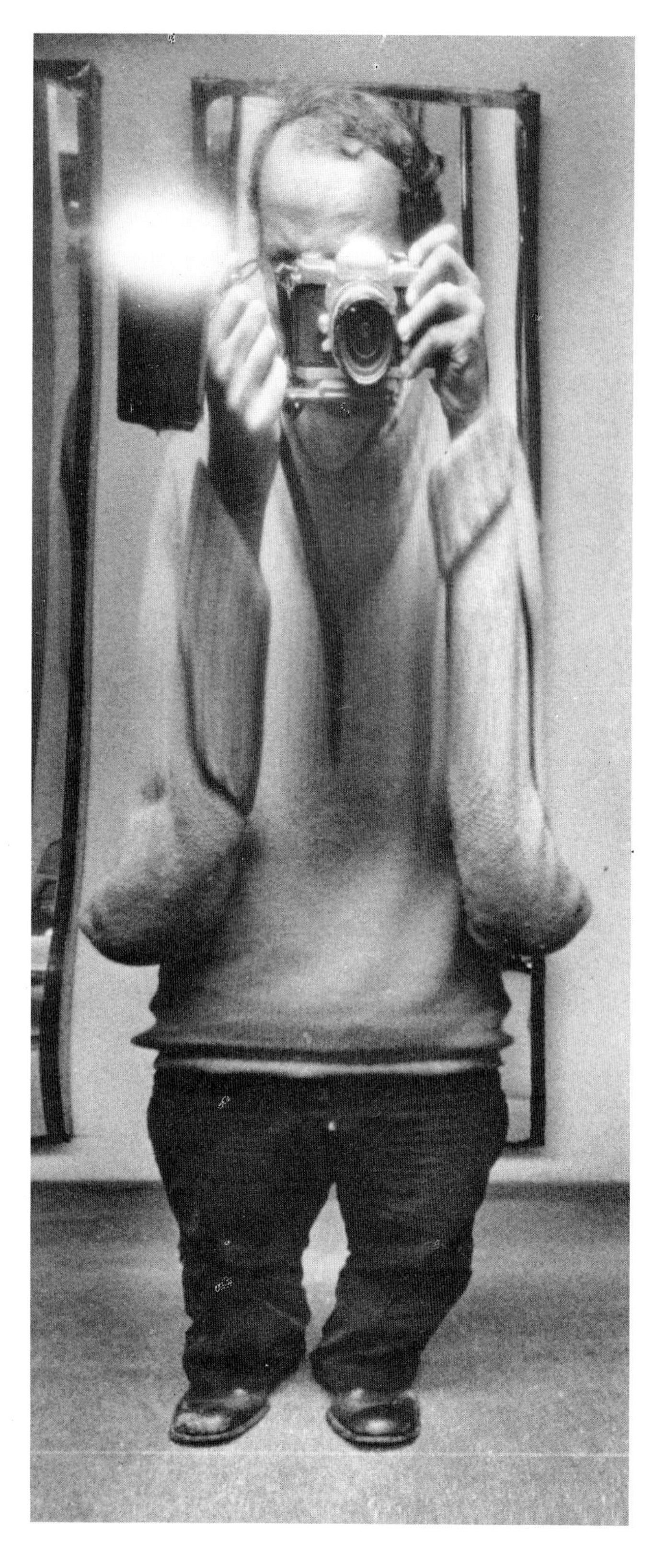

Georg Friedrich Bernhard Riemann (1826–1866) introduced topology under the name of analysis situs (analysis of position) as a modern branch of mathematics. Other contributors were A.F. Möbius (p. 73), C. Jordan, L. Kronecker, G. Cantor (p. 102) and H. Poincaré (who called it combinatory topology). This branch studies the properties of curves and surfaces that do not change under a continuous transformation. Two figures are topologically equivalent if we can deform one into the other continuously. Thus a circle is topologically equivalent to an ellipse, but not to a straight line or a circular strip. A sphere is equivalent to any convex surface, but not to a torus (p. 67), which has a hole. Topology has been applied to such famous problems as the Möbius ring—the one-sided strip (p. 67)—and the four-colour problem (pp. 77–82) as well as to the theorems of Euler and Jordan. Today topology is accepted as an independent subject and a basic structure of modern mathematics. These pictures reveal some topological distortions.
Right: horizontal distortion.
Left: vertical distortion.
Above: A cat's face distorted through a glass of water.

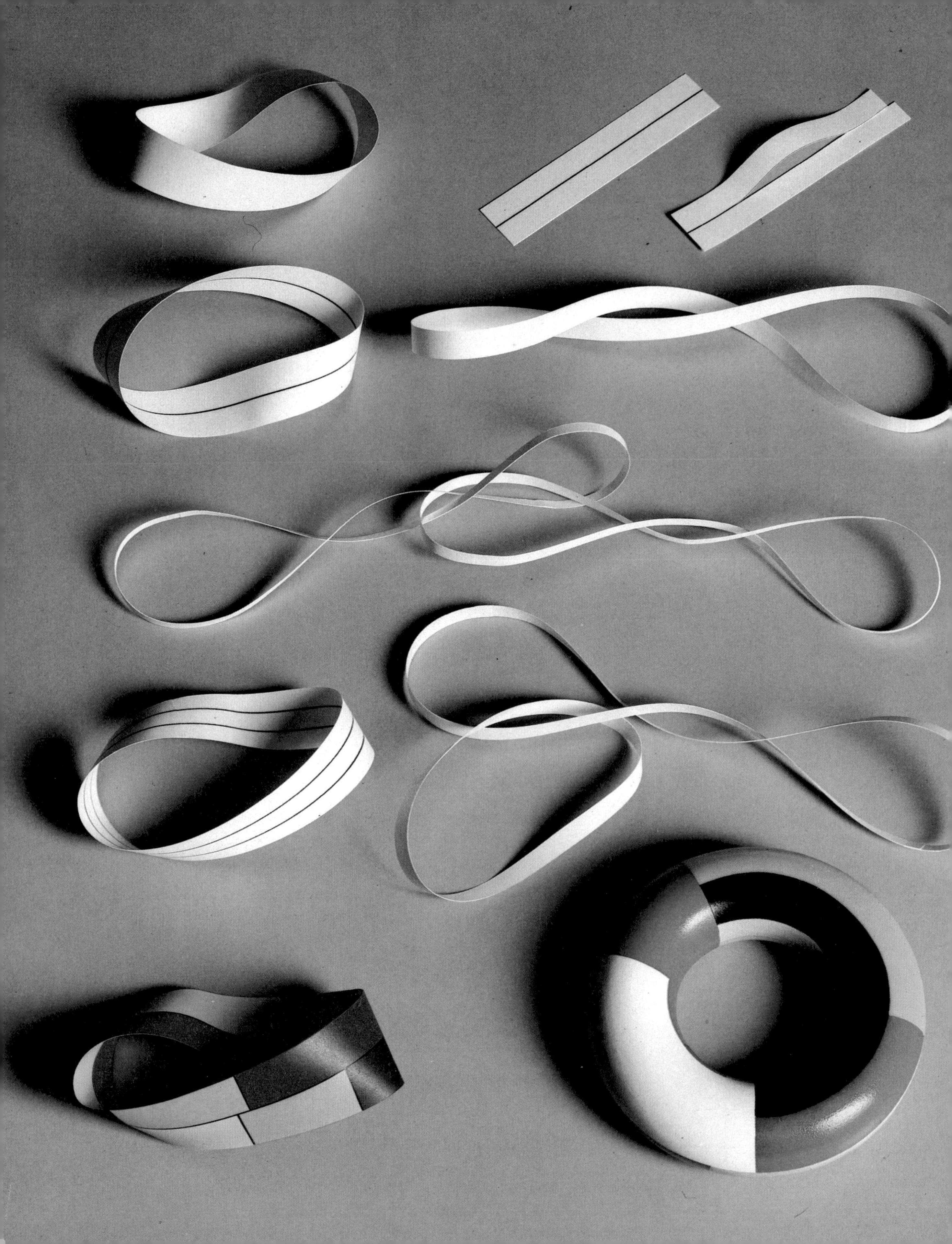

Opposite: Some interesting possibilities of one-sided strips: We usually think of a surface as having two sides. With a sheet of paper, for instance, we must round an edge to get from the front to the back. The Möbius strip, however, is an example of a one-sided surface.
Right: You can make a Möbius strip as follows: 1) Take a strip, twist it once and glue the ends together. Now, because the strip has only one surface and no edge need be rounded, any point can be reached from any other with a continuous line. 2) Next, cut the strip along its middle: it produces one loop with two surfaces. If you cut this again along its middle you get 3) two linked loops, each with two sides. 4) Take a Möbius ring longitudinally marked into three by two lines and cut the ring along the lines. It produces two linked loops, one a Möbius ring and the other a normal ring. In 5) there is a coloured Möbius ring. On a one-sided surface we might need up to six colours to draw a map (for a plane map, we have the four-colour theorem, cf. pp. 77–82). In 6) we have a coloured torus, a topological figure resembling a doughnut; any map could be drawn on its surface using seven colours.

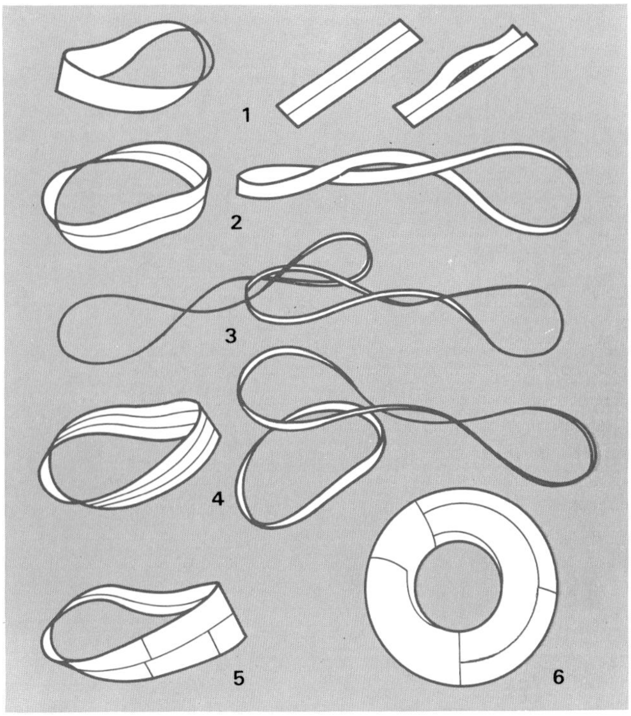

1
2
3
4
5
6

fig. 78

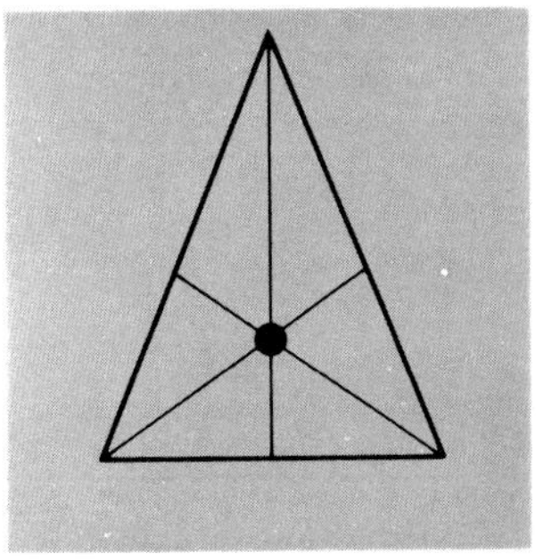

fig. 79

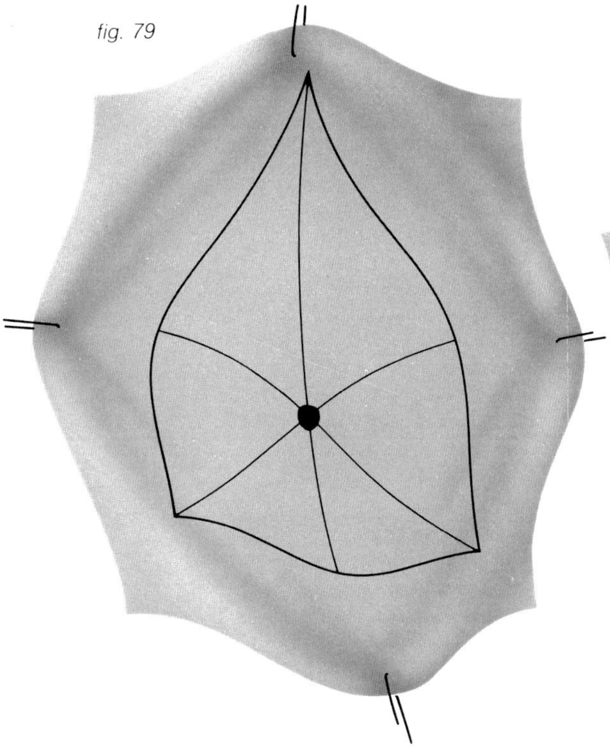

fig. 80

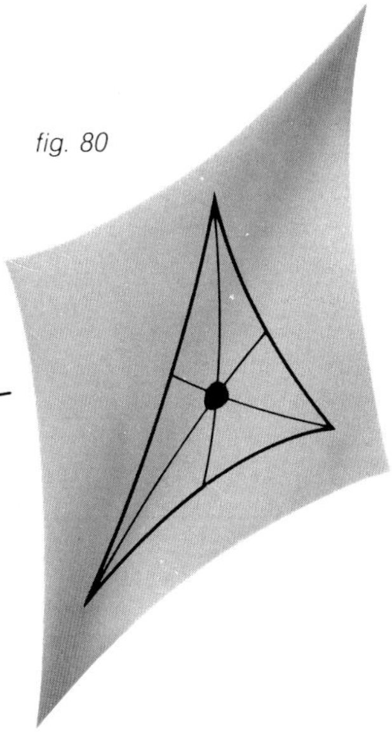

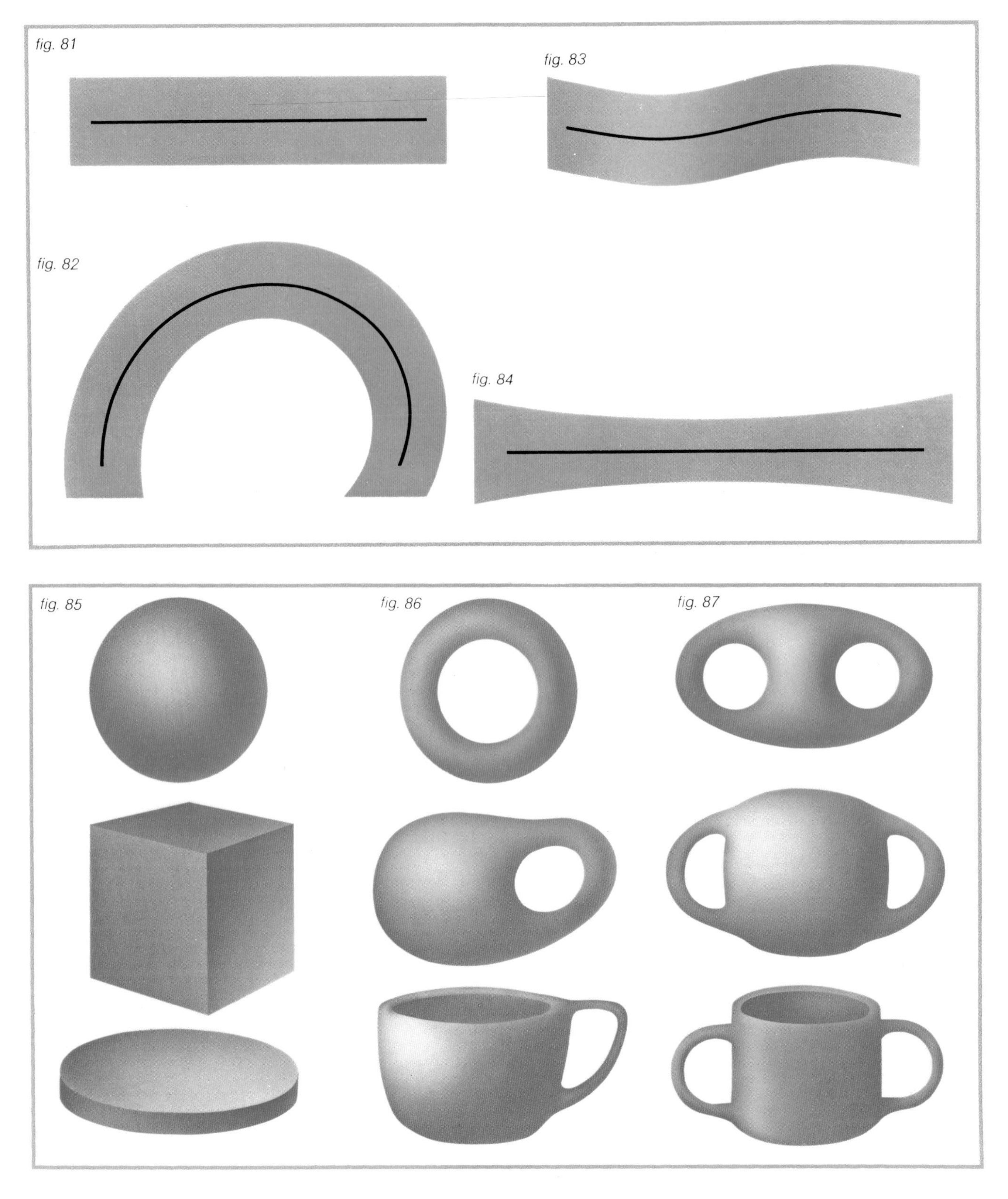
fig. 81
fig. 83
fig. 82
fig. 84
fig. 85
fig. 86
fig. 87

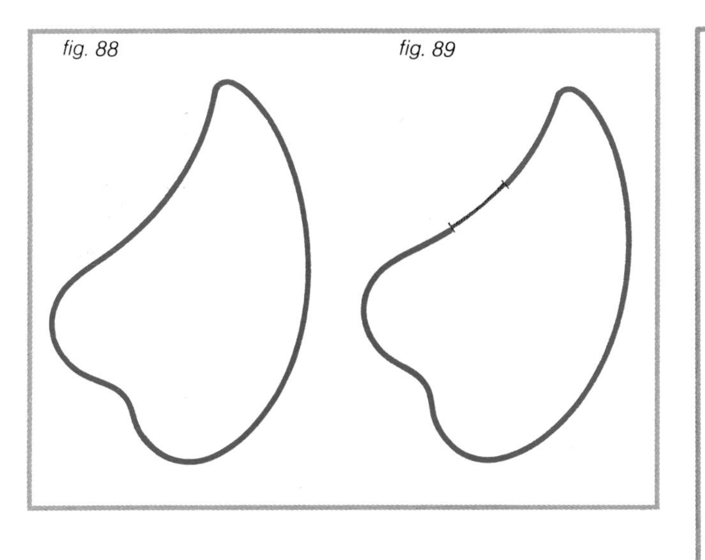

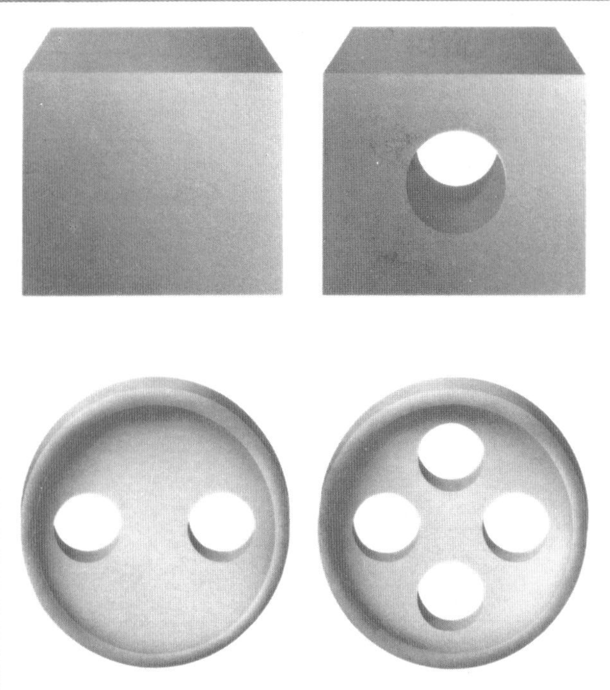

Right: A set of four topological objects with surfaces of genus 0, 1, 2, 4 (depending on the number of "cuts" needed to transform the body into one of genus 0, equivalent to the cube).

ularly interesting field of modern mathematics. The word topology means "study of place or position." What about geometry? It too studies points, lines and figures. True, but topology does not deal with the objects of traditional geometry. Topological objects can change their size and shape, be curved, twisted, squashed and generally deformed. Sometimes the objects are surfaces that cannot be constructed, or forms that seem unimaginable—pieces of paper with only one side for example. Some people have even dubbed topology "geometry on a rubber sheet." Indeed, topology studies those geometrical figures that retain their mathematical properties even when their size and shape change. Hence it is quite different from traditional geometry which is basically Euclidean, meaning objects do not change their size or shape. In topology we begin with the assumption that geometrical objects are not rigid and can undergo changes in size and shape when displaced.

Imagine a fairly elastic rubber sheet on which an isosceles triangle is drawn with a point marked at the intersection of its angular bisectors (Fig. 78). Let the sheet be stretched as seen in Figs. 79–80. It is not the fixed distances or angles (which can be altered by distortion) that concern topology, but rather the relative positions among points, lines and surfaces. (The positions do not change, even when the size and shape do.)

Take another elastic sheet and draw a straight line on it (Fig. 81). Under distortion the line becomes a curve (Fig. 82), or a wave (Fig. 83), or a lengthened line (Fig. 84), but it always remains a connected line that does not intersect itself.

Topology was systematically developed only in the last hundred years, although some earlier discoveries, chiefly those of Descartes and Euler, have since been incorporated in it. For further understanding of topological properties, consider some topological transformations of three-dimensional objects. Suppose we make a clay model of a sphere (Fig. 85), and transform it first into a cube and then into a slab. Next, take a ring (Fig. 86) and notice how it can be transformed into the handle of a cup. In Fig. 87 the double ring becomes a two-handed cup. What these three sequences share is that in each case, the trans-

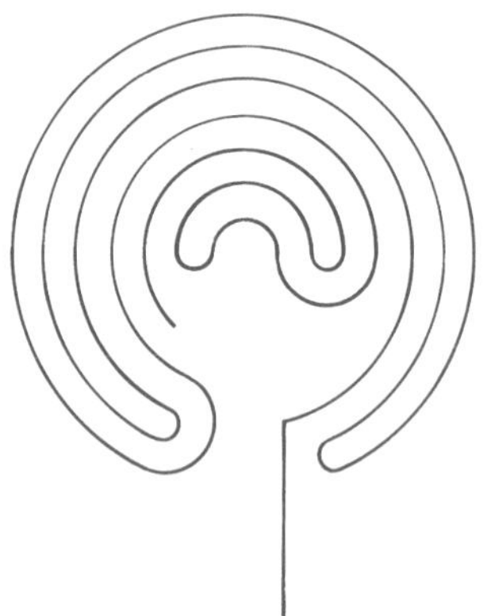

Left: The labyrinth at Cnossus, linked with the legend of Theseus, Ariadne and the Minotaur.
Top: Cretan maze with seven turns.
Below: The corresponding "thread of Ariadne" to be followed to reach the center.
Right: A 15th-century German maze from a plate by Johann Neudörffer the Elder (Nuremberg 1497–1563). Mazes have endlessly fascinated man who has invested them with a variety of meanings. They have inspired building decoration and ceilings; indeed the gardens of the nobility were often designed as mazes with networks of paths and avenues.
Opposite right: Four antique maze patterns used in French architecture.

formations preserve a property topologists call the genus of a surface. The genus is given by the number of holes in a figure that remain constant throughout the transformations. More technically, we speak of the number of nonintersecting and closed, or completely circular, cuts that can be made on the surface without cutting it in two. The three objects of Figs. 85, 86 and 87 are topologically different. The topological transformations occurring in each change the size and shape of the figure without producing any new topological feature. However, should we make cuts or produce new holes or tear a surface we would create new positional features. In a topological transformation we must not tear, cut or make holes. In Fig. 88 we have a closed curve; if we cut away a piece, the transformation is not topological, for we form a new figure equivalent to an infinite number of others, and all represented by an open line (Fig. 89). In topology, a square is equal to a triangle and a triangle to a circumference; all three are figures with an inside and an outside, and to pass from one to the other we must intersect the line.

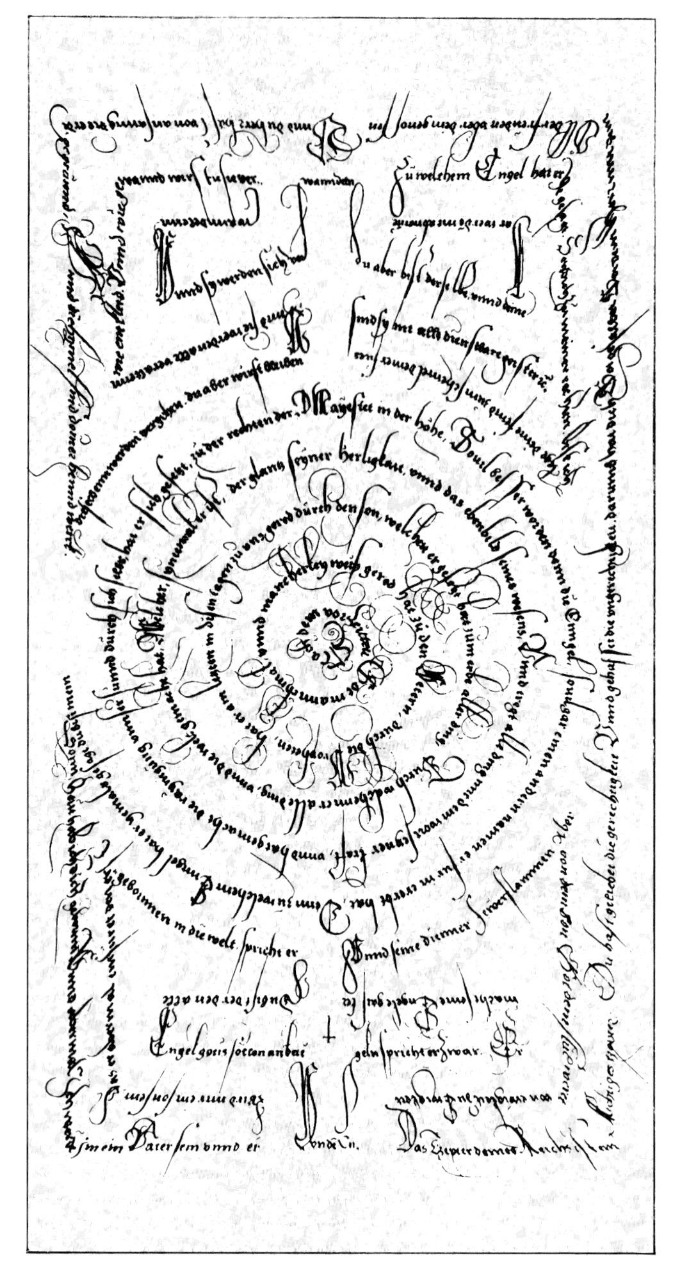

Topological labyrinths

The idea of inside and outside leads us to a series of geometric figures and mathematical problems as old as man: mazes. A famous maze of antiquity was the labyrinth of Cnossus on Crete which the mythical hero Theseus entered to kill the Minotaur, a monster with a man's body and a bull's head. The myth tells us that Theseus found his

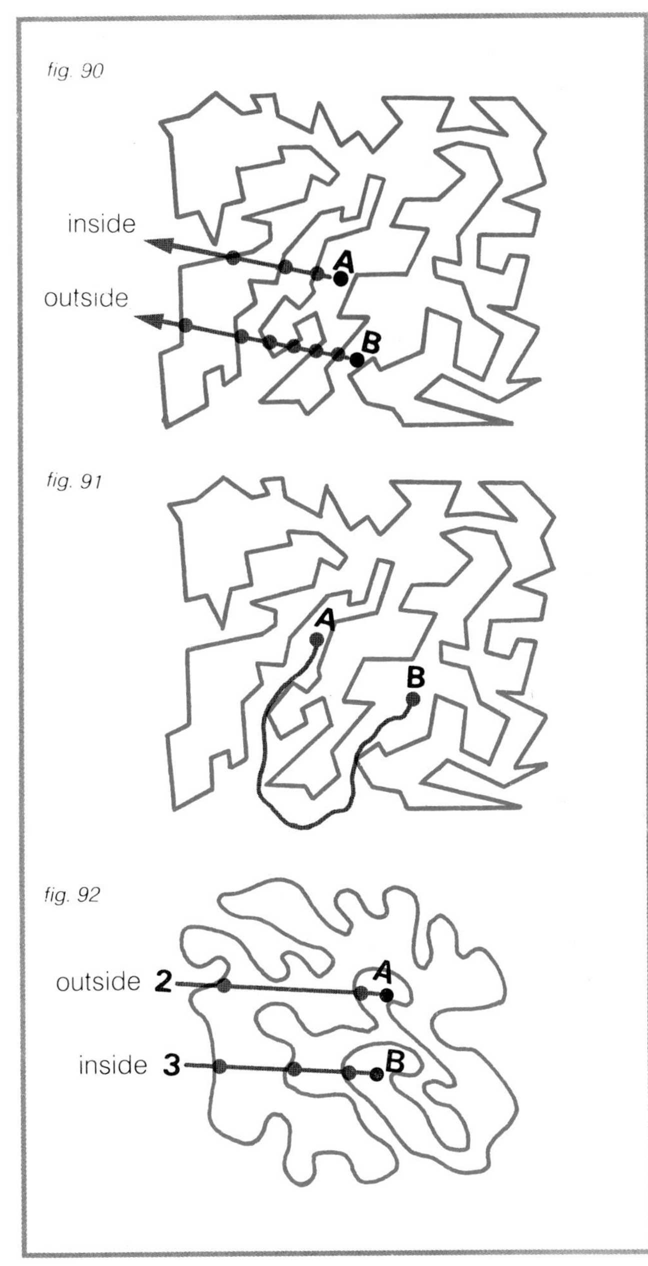

way out of the labyrinth by following the thread he unwound as he moved through the maze, a ruse conceived by Ariadne, who was in love with him. Labyrinths go back to the Neolithic period and the world's cultural history is replete with renditions of mazes, to be found in architecture, painting, literature, even film.

Broadly defined, mazes are geometric figures made up of lines which together form a topo-

logical plan. Such lines, with one outside opening, lead through a series of convolutions to a center, from which the outside is regained by returning the same way. Topological mazes, however, are not true labyrinths, but generally closed lines with an inside and an outside, not interlaced, but rather consisting of a set of curves, or a series of cornered contours of varying widths. These mazes are closer to a circle than to a genuine labyrinth. Like a circle, they are closed lines with an outside and an inside, and concern problems where it is difficult to establish, graphically or intuitively, whether a point is internal or external to them.

With a closed line such as that in Fig. 90, let us try to determine whether the points *A* and *B* are internal or external. There is a quick way to decide this if we know Jordan's theorem. Camille Jordan (1838–1922), a French mathematician, studied these problems and published his theorem in *Cours d'Analyse* (1882); if one traces a half line from each point to the outside zone and if the number of times the half line cuts the contours is even, then the point is external. If the number is

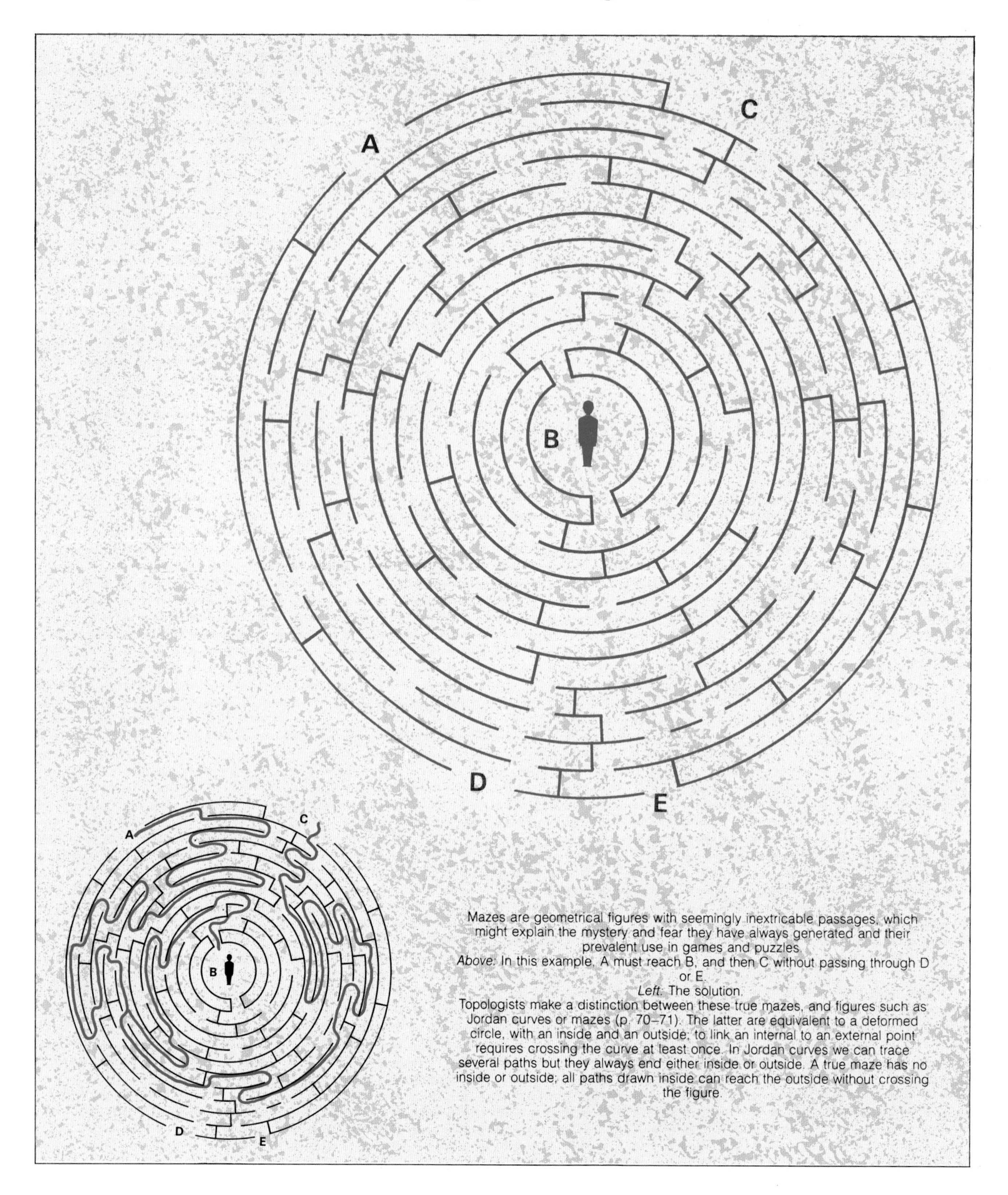

Mazes are geometrical figures with seemingly inextricable passages, which might explain the mystery and fear they have always generated and their prevalent use in games and puzzles.

Above: In this example, A must reach B, and then C without passing through D or E.

Left: The solution.

Topologists make a distinction between these true mazes, and figures such as Jordan curves or mazes (p. 70–71). The latter are equivalent to a deformed circle, with an inside and an outside; to link an internal to an external point requires crossing the curve at least once. In Jordan curves we can trace several paths but they always end either inside or outside. A true maze has no inside or outside; all paths drawn inside can reach the outside without crossing the figure.

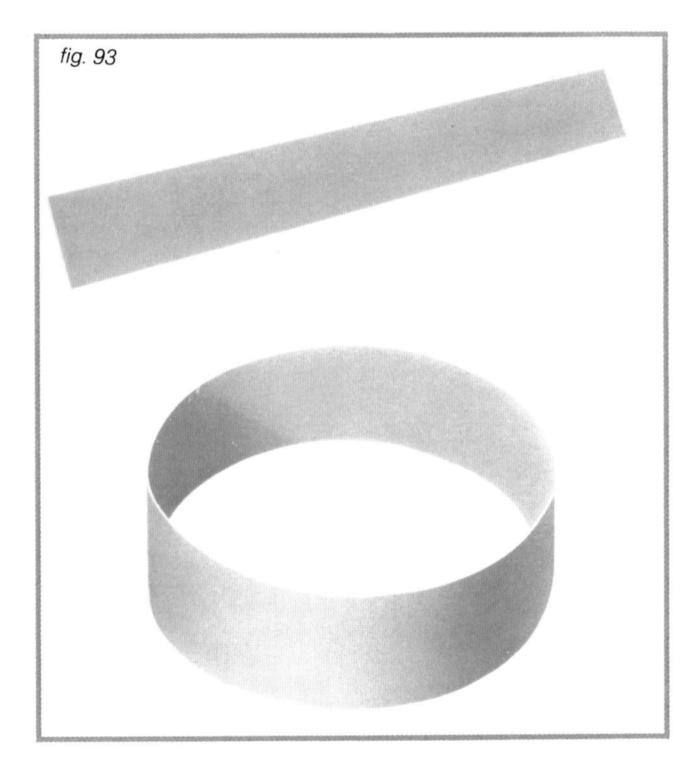

fig. 93

fig. 94

odd, the point is internal.

Jordan's theorem is a springboard for some amusing games. Take a closed line, as in Fig. 91, and ask someone to choose a point not on the line. Carefully observing whether the point is internal or external, we can show the person that we can pick another point in the maze and then join it to his with a line that does not cut the maze. We need only count an odd or even number of intersections according to whether the intersections on the line from his point to the outside were odd or even, and then draw the joining line. In Fig. 91, *A* and *B* are both external so there is a solution. In Fig. 92, however, one point is external and the other internal, hence no line can be drawn without intersecting the maze.

The Möbius ring

Among the founders of topology was the German mathematician and astronomer Augustus Ferdinand Möbius (1790–1868), who discovered a strange topological figure which was subsequently named after him. In a posthumous article he described the figure as a "strip without a second side."

Ordinarily we expect a surface to have two sides. A sheet of paper, for instance, has a front and a back, as does any other plane surface. However, Möbius managed to construct a one-sided strip; we cannot distinguish front from back, or an upper side from a lower side. To illustrate this visually, take a rectangular strip (Fig. 93) and glue the ends together to form a ring. It will have an inner and an outer side. If, at a given point, we begin to paint the outside green we will soon find that the entire outside is green. Similarly, we can paint the inside red. In a Möbius ring the result would be quite different as the two colours would overlap. Let us construct such a ring. We start with the same kind of strip but, before closing it into a ring, we give one of the ends a half-turn. We produce the ring of Fig. 94. If we now start colouring a side, as before, we find there is nothing left unpainted. An insect can walk along the strip and reach any point without crossing the edges.

Such properties are called invariant. They concern the single side and single edge of the

Escher's. *Möbius Strip II*. A 1963 woodcut, clearly inspired by topology (© Beeldrecht, Amsterdam 1982)

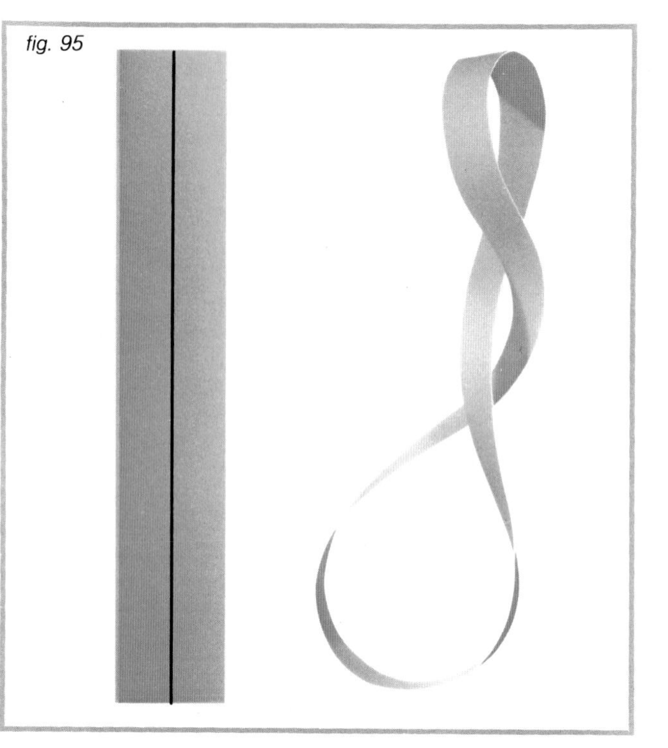

fig. 95

Möbius ring. They too suggest some entertaining tricks. Take two rectangular paper strips, equal in length and width, and have someone mark a point *A* on one side and *B* on the other; join the ends to make an ordinary ring of two sides. It would be impossible now to draw a line joining the points *A* and *B* without perforating the paper or crossing the edges. However, if the same points are marked on the second strip and this is glued into a Möbius ring, the points can then be joined.

Another surprising feature is that if we cut the Möbius strip lengthwise, we obtain a single twisted ring of two sides (Fig. 95), while an ordinary ring would divide into two separate rings. The explanation, though simple, is not immediately obvious, as we are not used to visualizing twisted rings. A Möbius ring has a single edge because its contour is a single closed curve; the cut merely adds a side. Unless one is familiar with these odd figures, it is hard to foresee the results. If we cut the twisted ring lengthwise a second time, there is a new surprise. We obtain two distinct but interlaced rings, and neither is single-faced (Fig. 96).

fig. 96

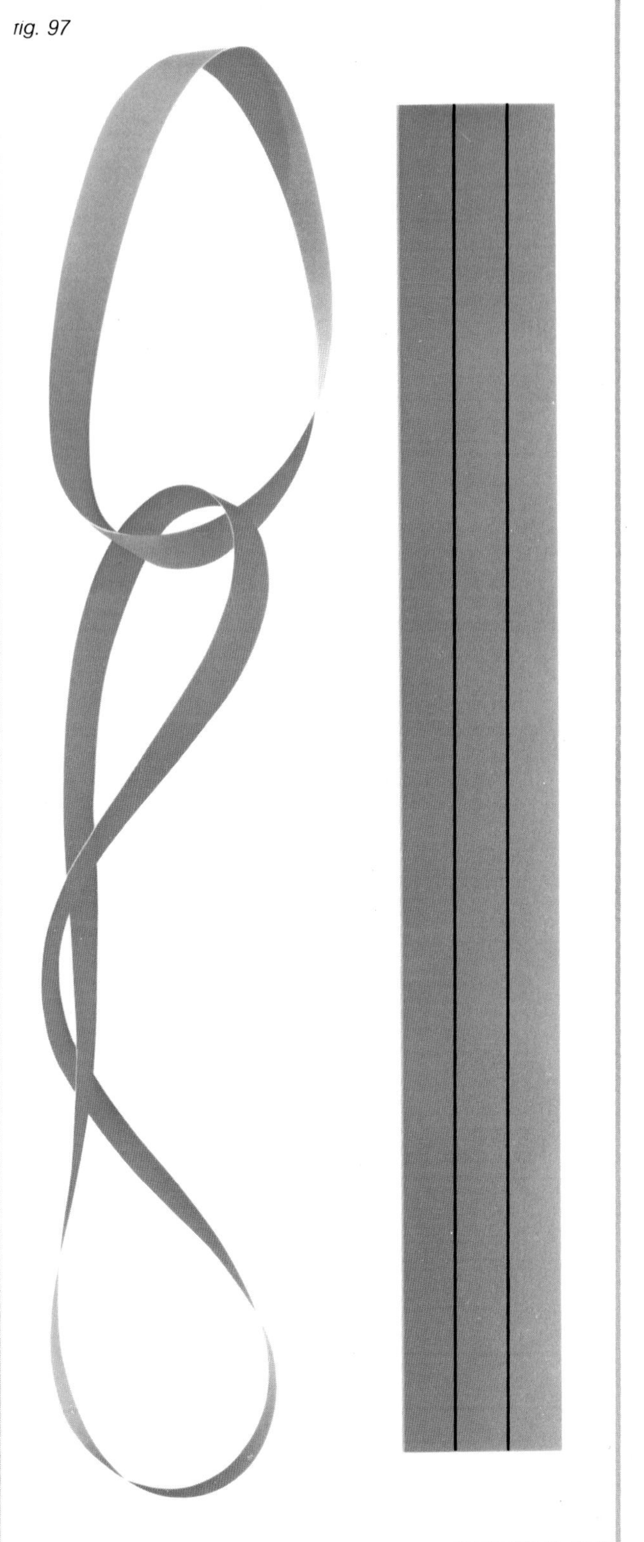
fig. 97

This last result might suggest that a Möbius strip must be cut twice to obtain two distinct and interlaced rings. Not so; this can be achieved with a single cut. Divide the original strip of paper lengthwise into three equal parts (Fig. 97) and shape it into a Möbius ring. Now a single cut starting on one line and running along both lines will produce two interlaced rings, one of which is a Möbius ring. This is a fascinating game and full of surprises if the Möbius ring is cut at varying distances from the edge.

Topological surfaces like the Möbius strip, originally just a mathematical curiosity, became increasingly important in both theory and practice. The formal features of such figures, for example, have served as models in physical research on subatomic particles. Recently an American industrialist used the theory to design a conveyor belt that is subject to wear on both sides rather than just one side. Such a belt will have twice the life of an ordinary one.

These examples illustrate the value of mathematical research. However abstract such research might seem, we should not question its

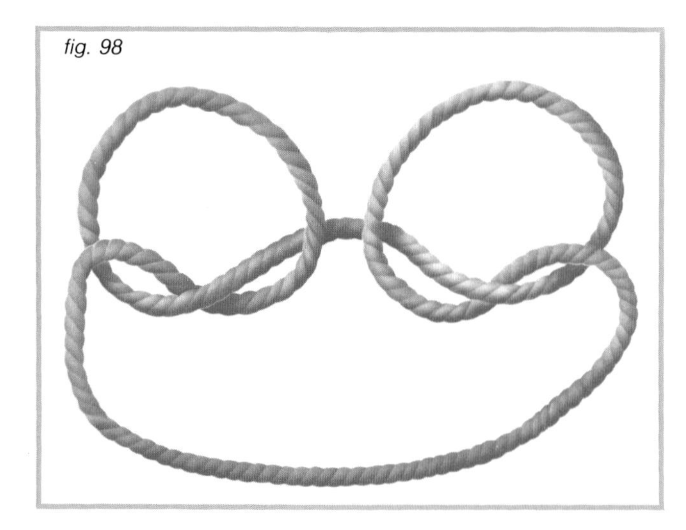
fig. 98

usefulness; even if at present it is of no direct practical use, in time it may contribute to the formation of new concepts which can be extended to other branches of mathematics and used by engineers to solve practical problems.

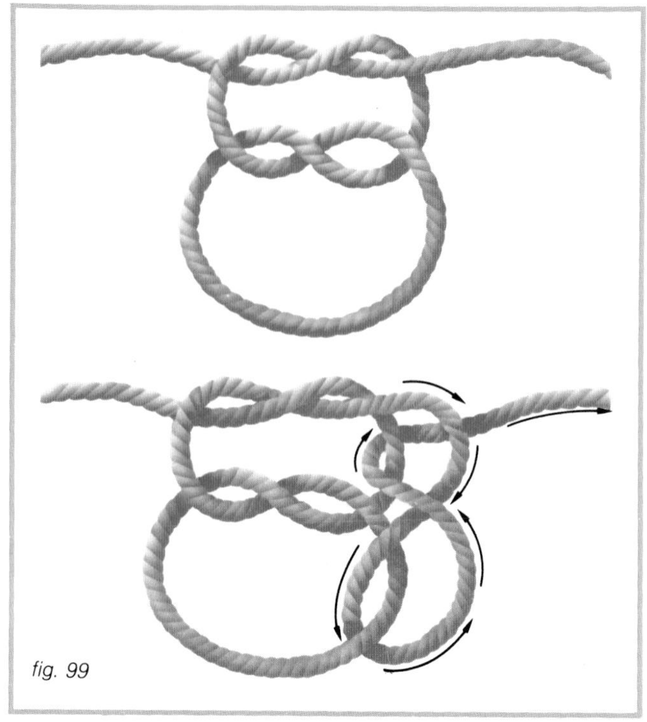
fig. 99

Games with topological knots

We all know how to tie a knot in a piece of string; you bend and loop the ends and pull them tight. If a knot is loose it can slip along a length of rope. Take the knots in Fig. 98 and consider their properties. They are opposed but cannot be undone. These are intuitive examples of topological knots of which a satisfactory theory has yet to be constructed.

Many magicians' tricks are based on the properties of knots, or rather false knots. One is illustrated in Fig. 99. The starting point is given at the top; one end is then passed through, as shown by the arrows at the bottom. If we now pull the ends the knot disappears. Let the reader try for himself. Although the subject of knots immediately recalls sailors' tasks or the out-door activities of boy scouts, knots provide an ample field of research for mathematicians. A simple knotted rope with the ends joined so the knot cannot undo itself is a good physical model for the concept of a mathematical knot. Part of topology is concerned with the theory of knots and examines the features of

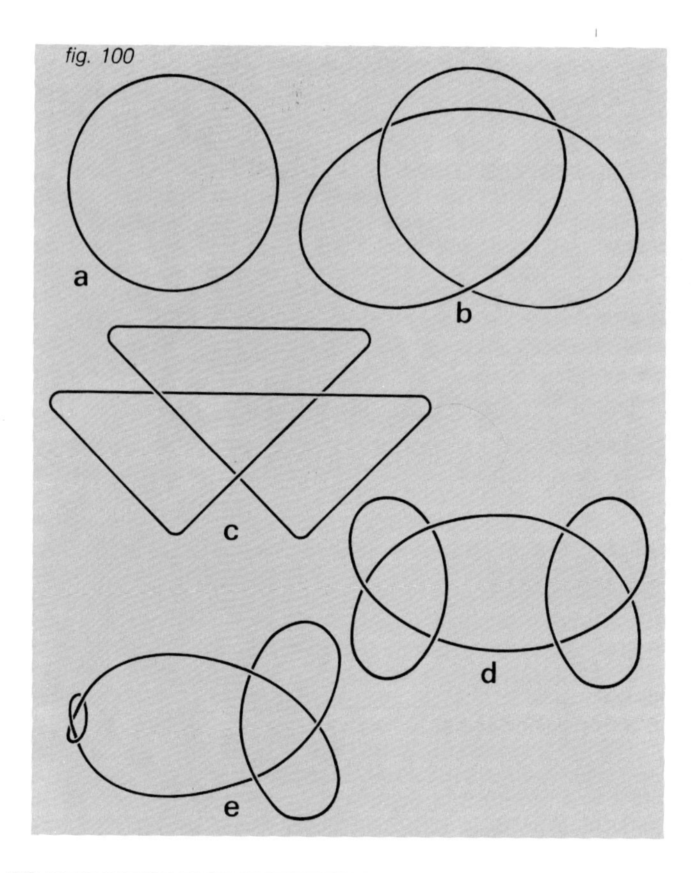

fig. 100

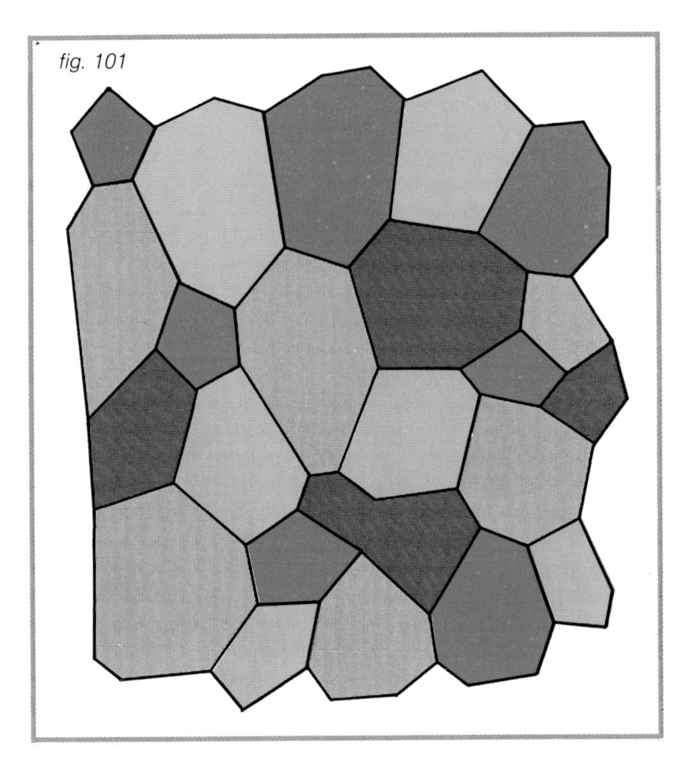
fig. 101

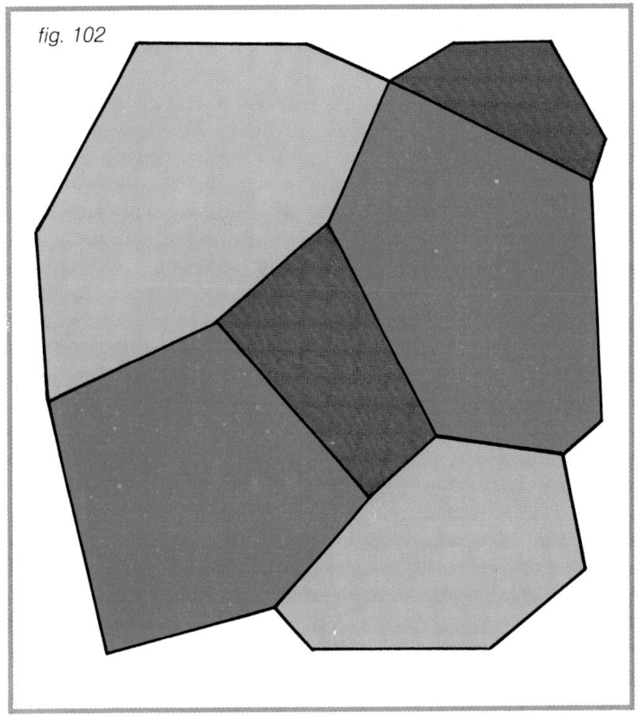
fig. 102

these geometric objects. How is a knot formally defined? In topology a knot is a one-dimensional curve starting and ending at the same point, and drawn in space without intersecting itself. The elements of Fig. 100 are all knots, including the circle which is a degenerate or trivial knot.

The theory of knots attempts to classify and analyze the mathematical features of such curves. Although we cannot examine its concepts and problems in depth, we can present a few intuitive notions. For example, the knots *b* and *c* differ only by the shape of their loops, while *d* and *e* differ only in size; hence, *b* and *c* should be considered equivalent as well as *d* and *e*. Defining the precise concept of equivalence is the fundamental problem of the theory.

Today, many mathematicians are studying ways of applying topological knots to various situations and problems.

The four-colour theorem

A familiar mathematical problem, with which mathematicians, amateurs and students have wrestled for many decades, is the four-colour problem. It was first formulated in 1852 by Francis Guthrie in a letter to his brother Frederick, a student of the noted mathematician Augustus de Morgan.

In drawing a geographical map with numerous states, it is customary to indicate adjacent countries with different colours. How many colours are needed to cope with any map? Over the years it has been found that however complex the map, four colours suffice; less will not serve in all cases. Take Fig. 101 and try with three colours. It cannot be done: four are needed.

In the particular case of Fig. 102, three colours are actually enough. However, the fact that four were deemed sufficient for any map, plane or spherical, led mathematicians to state the following theorem: *For any subdivision of the plane into regions that do not overlap, we can always show these regions with four different colours in such a way that no two contiguous regions are the same colour.* Regions are contiguous if they have a common boundary not reducible to a point. Thus the horizontal and vertical squares on a chess-

Right: The component parts of a Rubik's cube and some of its positions. 1) Starting position with each face uniform in colour. 2) Cross design produced by the sequence of moves outlined on pp. 89–90. 3) Random arrangement. 4) The skeleton of the cube with its six faces on six arms converging on a core. 5) Edge cubes, with two colours. 6) Corner cubes, with three colours. Rubik's cube (named after its inventor Ernö Rubik, an Hungarian architect) is of such wide renown that it occasioned a world championship, held in Hungary in 1982.

board are contiguous, diagonal ones are not (they touch only in a point).

Many 19th century mathematicians tried to solve this problem, but without success. In 1879, Alfred Bray Kempe, a London solicitor and member of the London Mathematical Society, published an essay in which he presented a proof of the four-colour theorem. Even though his proof was subsequently found to be defective, it was by developing Kempe's ideas that 97 years later a correct proof was given. The solution came in 1976, and relied heavily on the use of high-speed electronic processors. In the interim period, however, and particularly in the 1930s and 1950s, important discoveries were occurring in mathematical logic regarding the concept of proof. For example it was found that there are some theorems that can be stated fairly briefly, but their proofs are so lengthy it would take years to write them out completely. (Detailed examples are beyond our scope here, but the interested reader will find appropriate references in the bibliography.) With the advent of ultra-high-speed processors, problems connected with the four-colour theorem became solvable, and proof was given in June 1976 by Kenneth Appel and Wolfgang Haken, mathematicians at the University of Illinois, after using 1200 hours of computer time. In principle other mathematical problems can also be solved, but it would demand a computer as big as the universe working at least as long as the world is old (see Larry J. Stockmeyer and Ashok K. Chandra, "Intrinsically Difficult Problems," *Scientific American,* vol. 240, no. 5, May 1979).

fifteen
CUBO

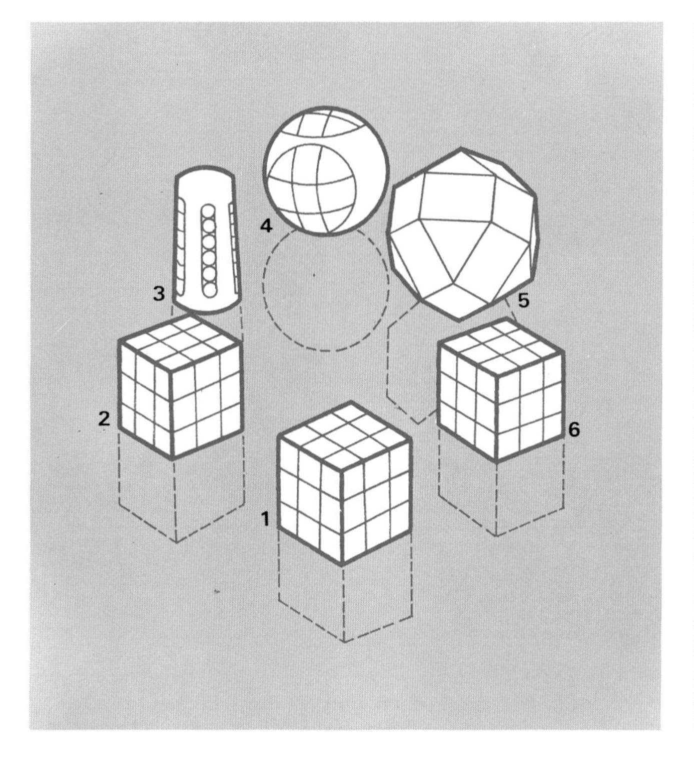

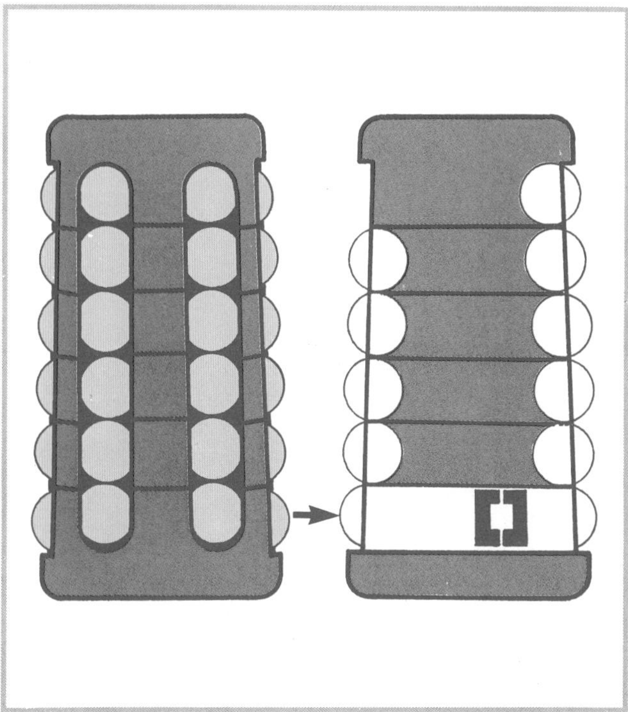

Left: A selection of games developed since the introduction of Rubik's cube.
1) The familiar Rubik's cube; 2) a cube with numbers instead of colours; 3) the tower of Babel; 4) a sphere with detachable and turnable parts modelled on Rubik's cube; 5) Rubik's magic snake; 6) a cube with split colour scheme.

Above right: The tower of Babel. At first sight this game with some 10^{30} combinations seems unrelated to Rubik's cube. It is about 3.5 inches high and has six columns, each containing six marbles in the colours blue, grey, red, brown, yellow, green. The marbles can move horizontally or vertically. The column is divided into sections which can be easily turned about the vertical axis, while the vertical movement is made possible by two openings in the base which can swallow a marble and thus lower the entire set above it. The game is to restore the original position after scrambling the marbles.

Right: Some of the constructions possible with the magic snake. More than 2000 figures can be formed in this new game from the cube's inventor, among them animals, birds and unusual geometrical patterns.

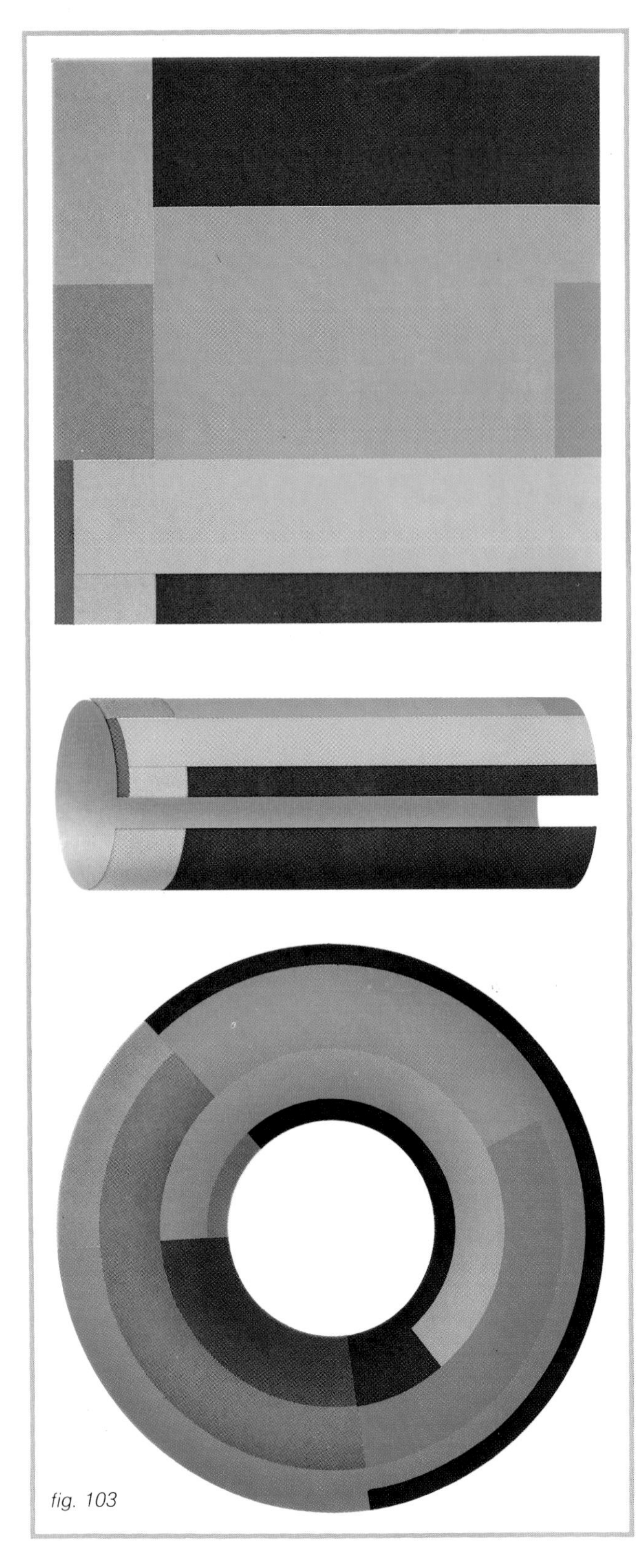

fig. 103

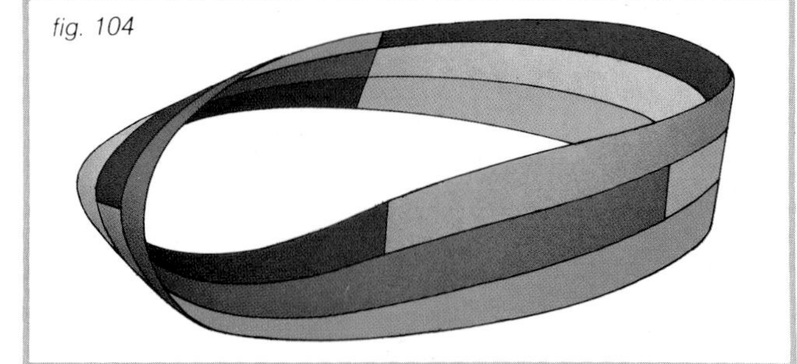

fig. 104

Opposite left: The successive figures illustrate a familiar topological problem.
Top: A rectilinear pattern with adjacent regions differentiated by only two colours.
Center: A random line is drawn. With only two colours, can we still differentiate between the regions?
Bottom: The solution. Invert the colours above the line.

Topologists and various amateurs amuse themselves with problems analogous to the one involved in the four-colour theorem. It may seem surprising but they have succeeded in proving similar theorems for more complicated surfaces, such as the Möbius strip or the torus (surface of a solid ring, Fig. 103).

Paradoxically, the analysis of simpler geometric surfaces is harder than that of more complicated surfaces. It has been proven that on a Möbius strip the map problem requires six colours (Fig. 104), while a torus requires seven colours. Starting from the four-colour theorem we can construct a very simple game for two people. Take at least five pencils of different colours and a sheet of paper. The first player traces a circle on the paper as in Fig. 105a. The second player colours the circle and traces a second adjacent circle (b). The first player shades the second circle with a different colour and then traces a third circle (c), and so on. The players continue, with the proviso that adjacent circles be of different colours and no more than four colours be used. The loser is the player forced to use a fifth colour because of his opponent's cunning in tracing adjacent circles.

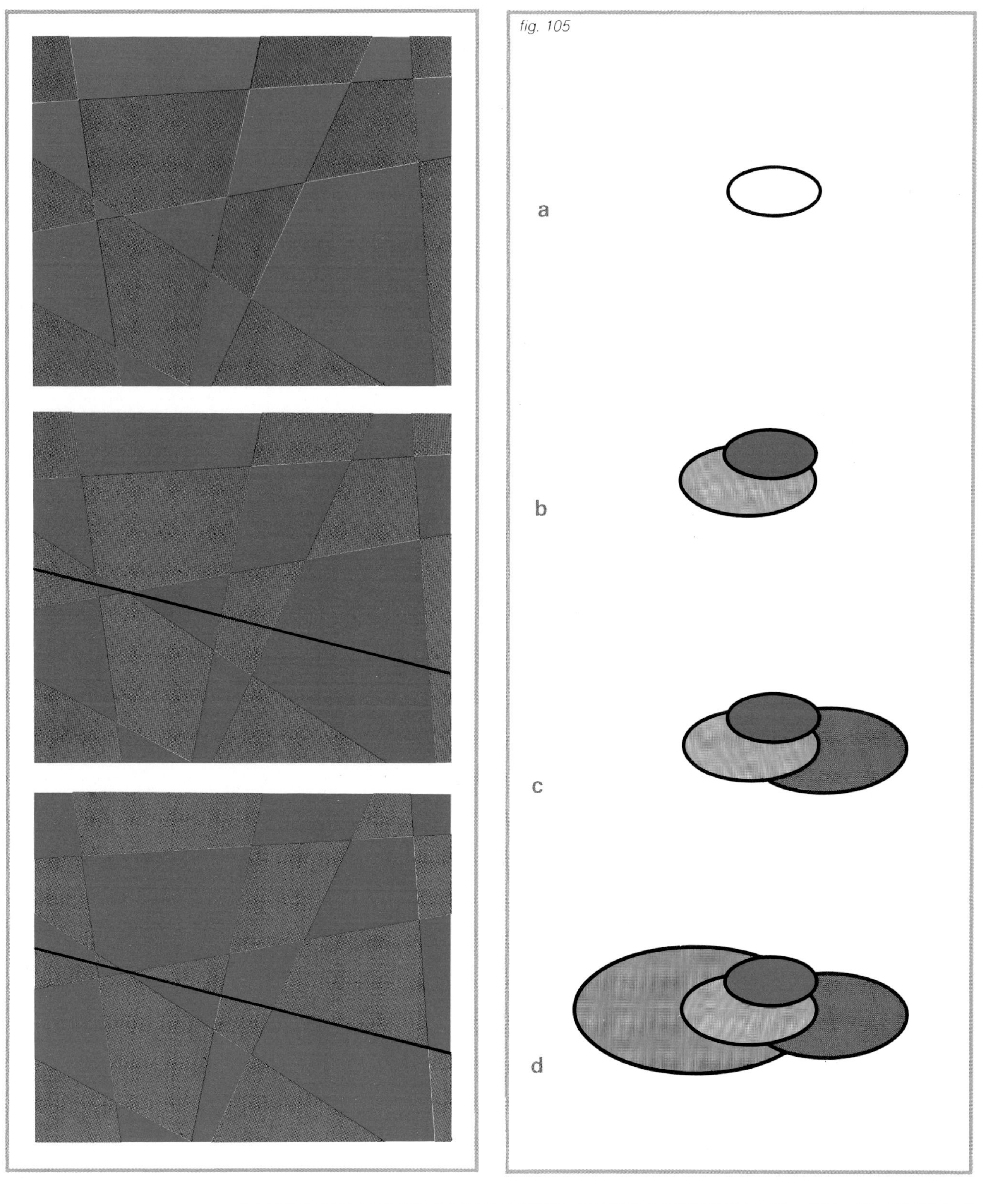
fig. 105
a
b
c
d

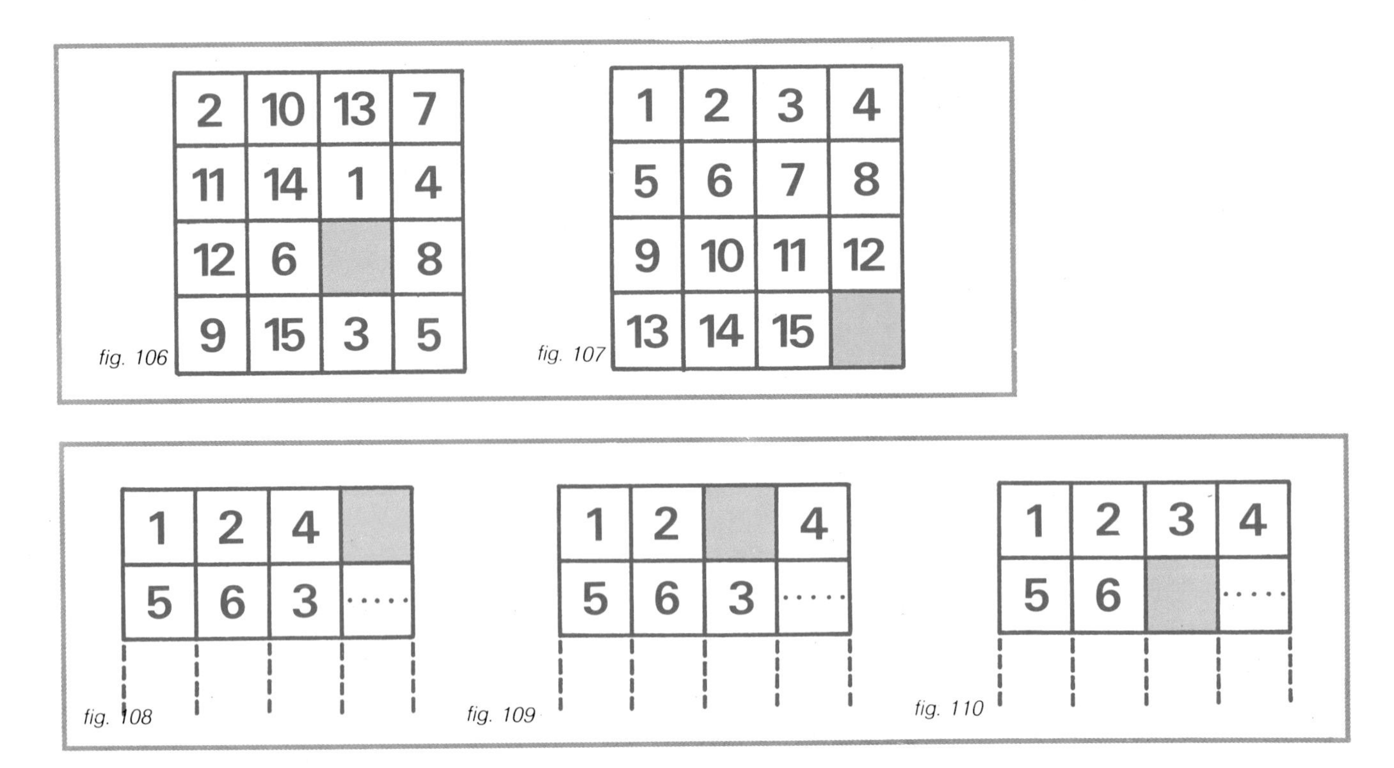

fig. 106 fig. 107 fig. 108 fig. 109 fig. 110

Rubik's cube

The inventor of this magic cube is the Hungarian Ernö Rubik, who describes the object as "an admirable example of the rigorous beauty and the great wealth of the laws of nature; it is a surprising example of the admirable capacity of the human mind to demonstrate the scientific rigour (of the laws of nature) and to master them . . . it is an example that shows the unity of the true and the beautiful which for me are one and the same." (See André Warusfel, *Réussir le Rubik's Cube,* Denoël, Paris 1981.)

This puzzle first appeared in England in 1978 and has since captured the imagination of mathematicians and puzzle enthusiasts around the world. Miniature cubes, blossoming as pendants, as tiepins and on key rings, are now all too familiar objects. Youngsters everywhere can be seen manipulating the multicoloured cubes to attain a certain combination. Indeed it is so popular that not only have millions of cubes been sold and numerous books and articles on its properties written, but in 1982 a Rubik's Cube world championship was held in Budapest, Hungary. There are even those who insist that "cubology" is a new science, in its early stages, to be sure, but destined for a grand future.

The Rubik cube consists of 26 small multicoloured cubes which together form a cube of $3^3 = 3 \times 3 \times 3$. The game consists of returning the coloured cubes, after they are mixed up, to their initial position so that only one colour is on each side of the cube.

A close relative

This is an appropriate moment to recall the Fifteen Puzzle, invented in the United States by Samuel Jones Loyd in 1870, as it is the two-dimensional forebear of the Rubik cube. It consists of a square 4×4 with 15 movable numbered square markers (Fig. 106). It has about 21 million million arrange-

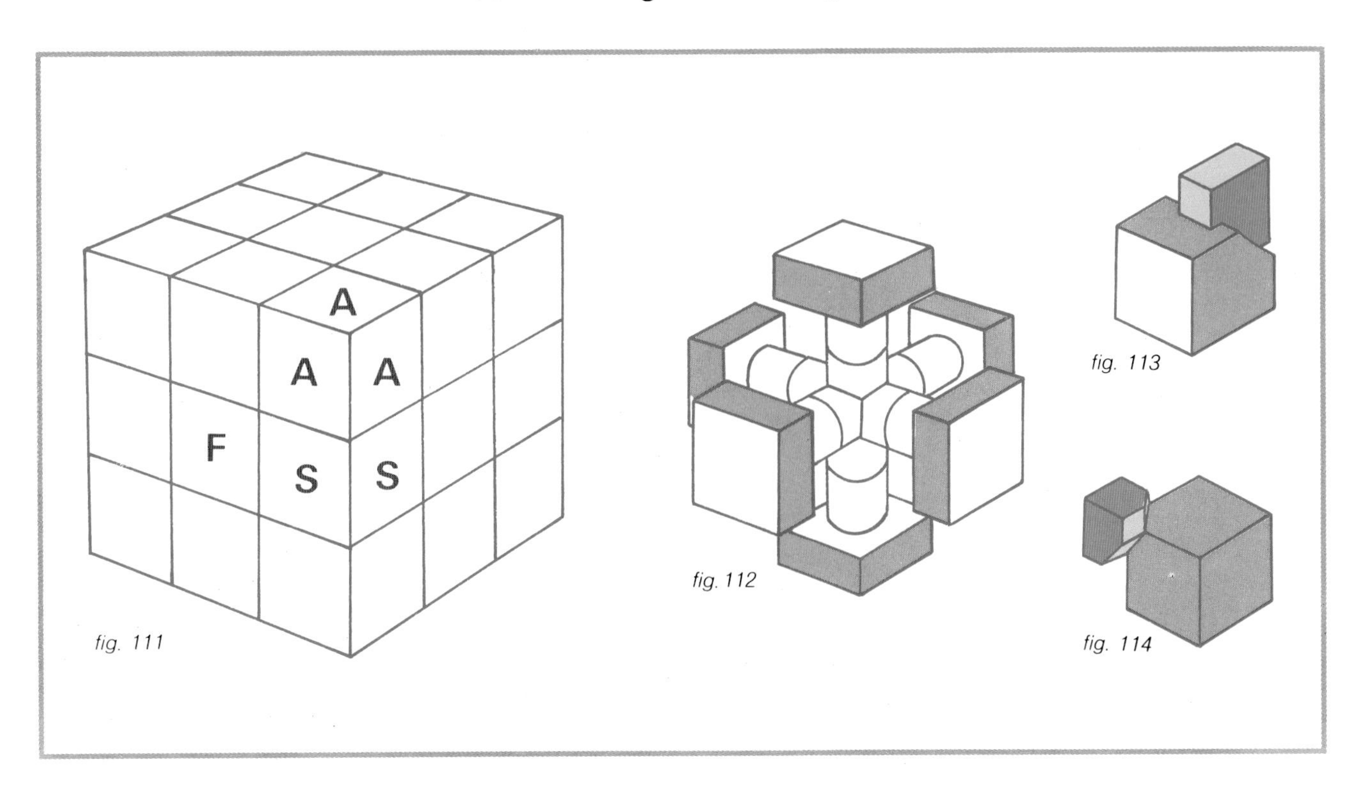

fig. 111

fig. 112

fig. 113

fig. 114

ments. Rubik's cube has over two million times more, namely 43,252,003,274,489,856,000; at the rate of one trial a second, it would take 1,360,000,000,000 years to check them all. A man's lifetime would be too short to even review them; indeed the vast number of possibilities in these games is their allure. The Fifteen Puzzle, although not as widespread as Rubik's cube, is nevertheless interesting and a study of it helps us to understand the cube.

On the 4 × 4 chessboard are 15 counters numbered 1 to 15. They occupy all but one square which is empty. The game is to start from a random position (Fig. 106) and restore order (Fig. 107) through a series of moves consisting of horizontal or vertical displacements into the empty square.

Here we confine ourselves to one stage of the game, and to the top two rows only. Suppose we have reached the position of Fig. 108. The next move will be to push the 4 into the gap and then the 3 up into the new gap, leading to Fig. 110.

Short history

The idea of the magic cube was developed independently in both Hungary and Japan. Ernö Rubik, an architect born in 1944 in Budapest where he now teaches planning and construction, was the first to present a study on the cube, which he did in the mid-1970s. In 1975 he patented the simple and ingenious mechanism that enables the small cubes to be rotated. Only about a year later a Japanese engineer, Terutoshi Ishige, discovered a similar idea and obtained a Japanese patent. In Hungary, where some two million cubes have been sold, it was used as a teaching aid in schools before being adopted as a game. Certainly the concentration required for even the simplest solutions is a worthy exercise in mental discipline. From the outset small competitions were held; a twelve-year-old Hungarian boy seems to have reached the correct position in a record time of twelve seconds.

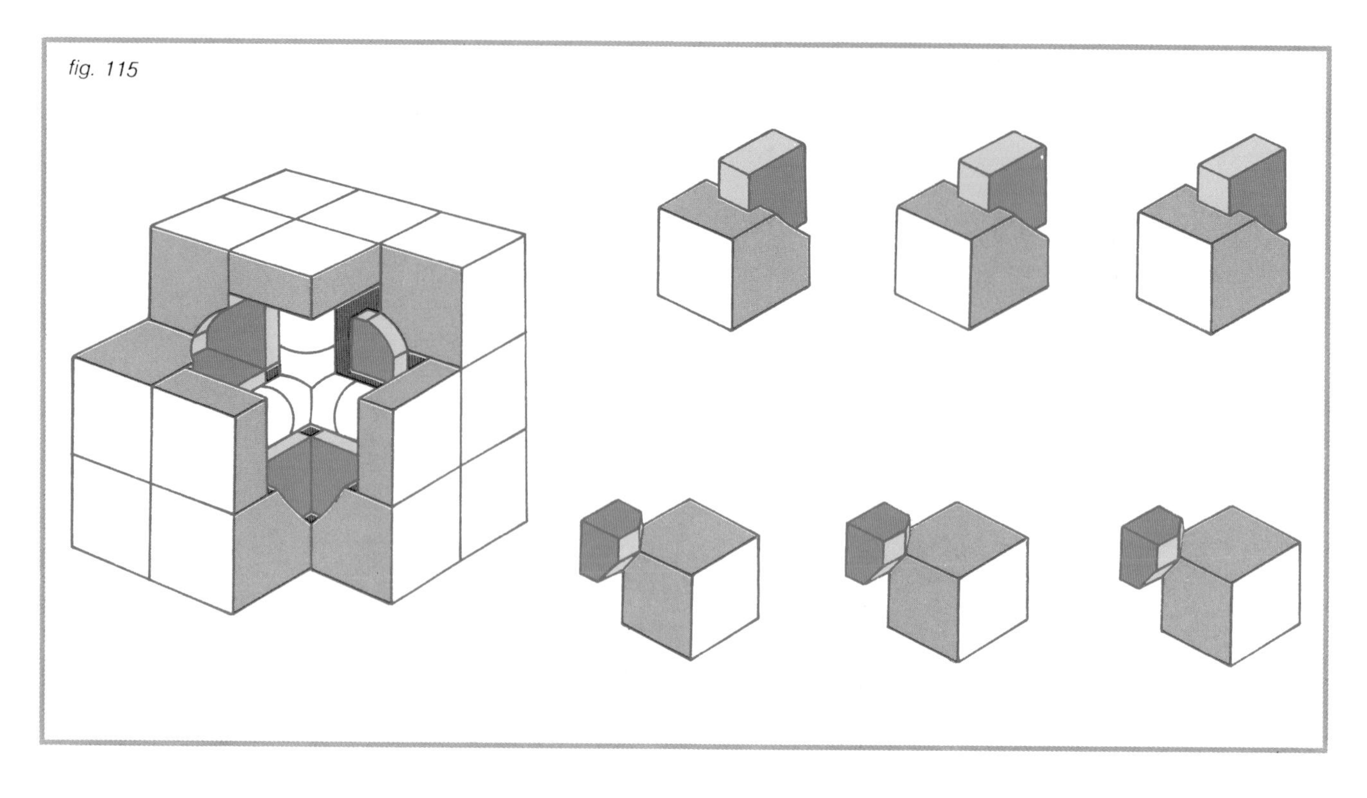
fig. 115

A rather simple mechanism

There is in fact a puzzle within the puzzle. Try to imagine the mechanism that makes the various rotations possible and then draw it. One is inclined to think of elastic materials, wires with specific properties, or perhaps a block held together by a magnet. However, the solution is far less complicated and purely mechanical. We will try to describe it even though an initial understanding really requires handling the cube. Three types of small cubes are used to form a Rubik's cube which has 8 cubes at the corners, 12 cubes along the edges, and 6 central cubes (Fig. 111). The central ones have a single face (Fig. 112), those along the edges have two faces (Fig. 113), while those at the corners have three faces (Fig. 114). Fig. 115 helps us to grasp the play of recesses that allows the faces to turn.

What makes the cube a unique puzzle are the games of restoring order. The order, or regular arrangement required, can take various forms. In general we start with the colours on the cubes mixed up and then try to reach certain regular patterns by making a series of moves. It is particularly difficult to return to the cube's original condition where each face is of uniform colour. As no one has yet succeeded in doing it merely by trial and error, we must proceed rationally by using the branch of mathematics, group theory, that deals with the various arrangements of the cube, and by creating a kind of science complete with symbols and rules.

By far the simplest way to restore the cube's initial state is to dismantle it first. Give the cube's top layer an eighth of a turn and insert a lever (the handle of a spoon, a thin key, or some such object) between the top face and the black triangle exposed (Fig. 116). Lift, as shown by the arrow, and the little cube along the edge will leave its socket, revealing part of the internal mechanism (Fig. 117). A corner cube can now be removed by turning it inward; the two cubes immediately under it—one edge and one corner—will then slide off easily leading to Fig. 118. If the operator places the extracted cubes with their coloured

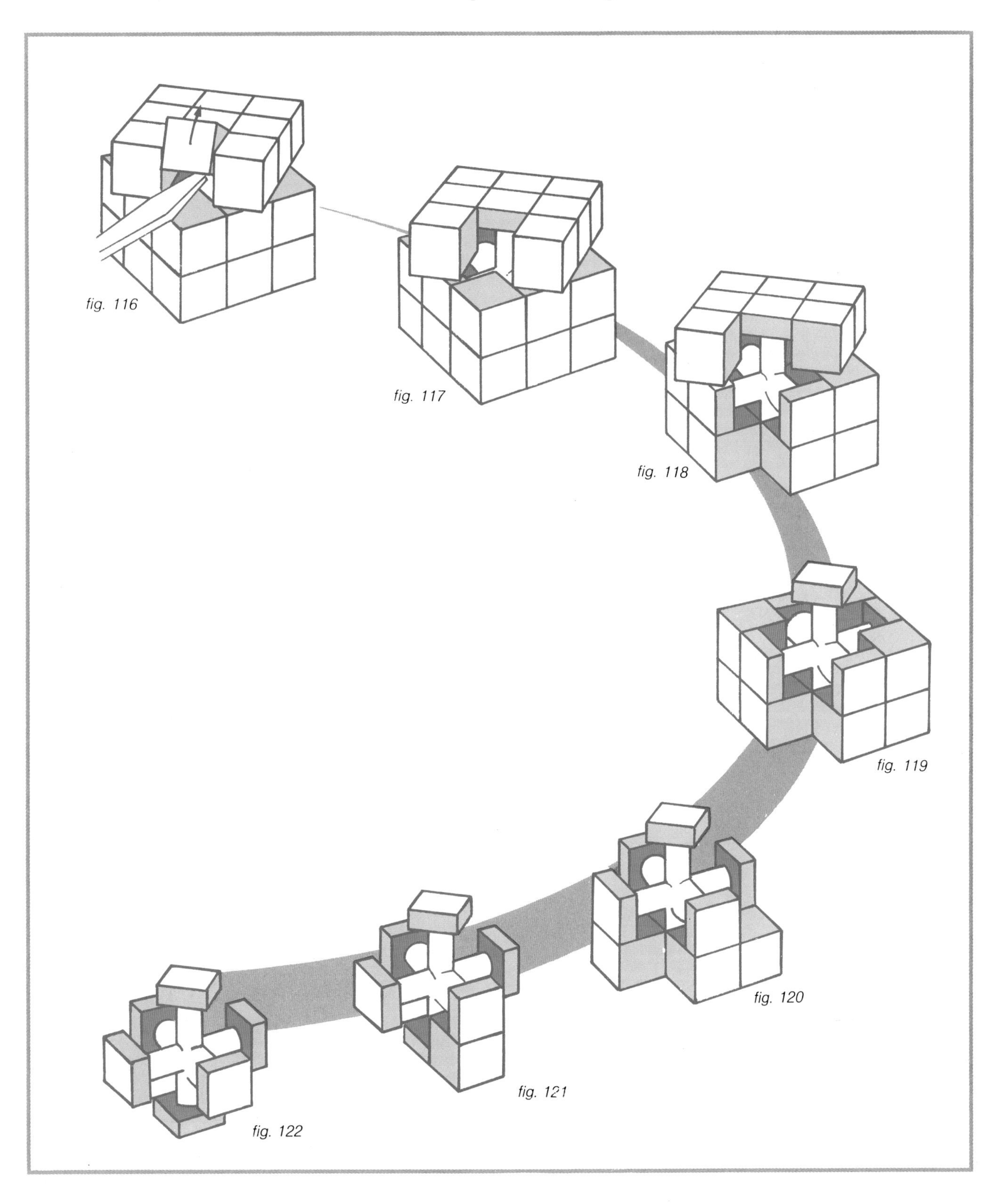

fig. 116

fig. 117

fig. 118

fig. 119

fig. 120

fig. 121

fig. 122

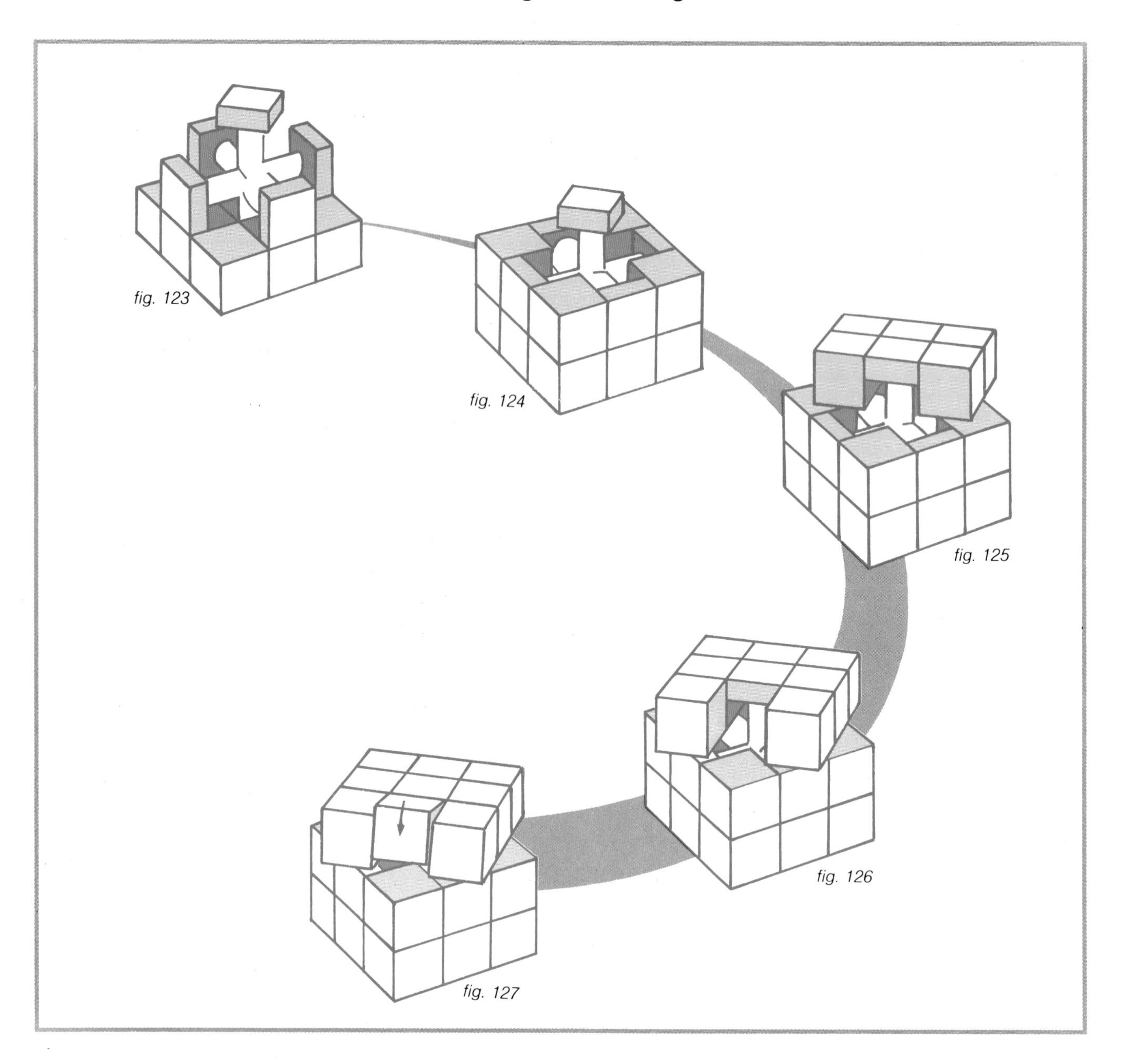

parts facing him, he will make subsequent reassembly easier. Next, extract the entire top layer by disengaging the pieces on the side from which a corner and edge cube have been removed. This leads to Fig. 119.

Once the top layer is off, the other cubes will readily peel off and should be dislodged from top to bottom (Fig. 120). The bottom layer comes off almost by itself (Fig. 121). In the end only the skeleton remains with its six fixed facets of different colours—not at all cubic in shape—attached to the three orthogonal axes converging toward a pivot of six arms placed at the center (Fig. 122). The three axes can turn only on themselves. Unlike these six facets, the other pieces are proper cubes, each with a kind of rounded foot pointed to the center of the magic cube and with recesses toward the inside. It is now simple to grasp the

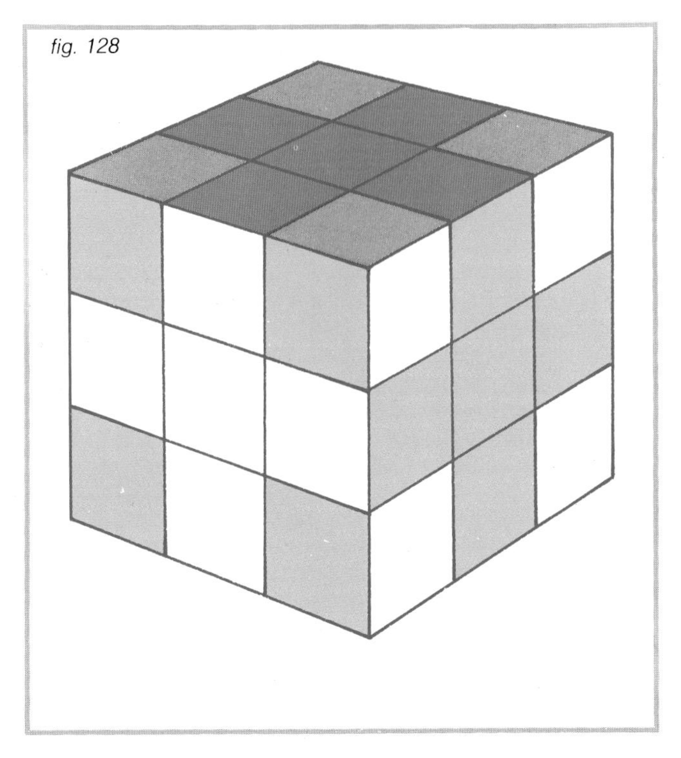
fig. 128

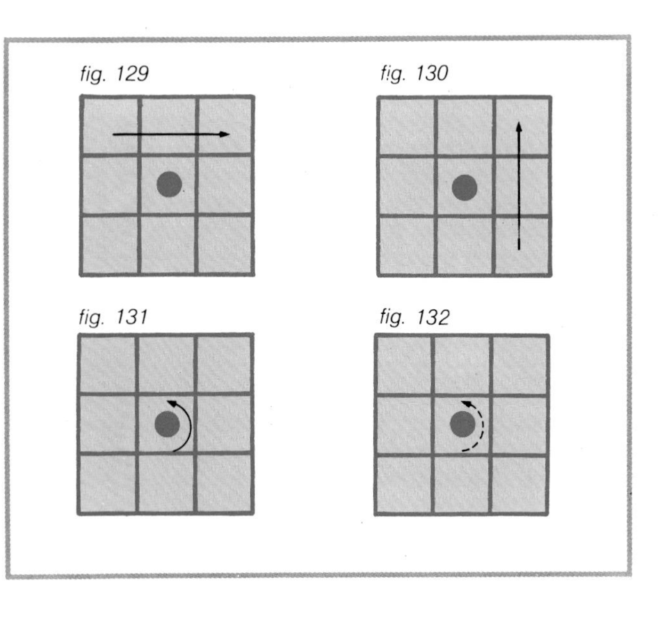
fig. 129 fig. 130 fig. 131 fig. 132

rotatory mechanism, which depends on the cubes supporting each other by their feet without being mutually attached. Thus the edge cubes hold the feet of the corner cubes; these in turn support the feet of the edge cubes. The keystones are the central cubes. We can now reassemble the cube. First, we choose a colour. Suppose we start with orange for the bottom face. We then take the edge cubes orange-white, orange-green, orange-yellow, and orange-blue. Sometimes a slight push or twist is needed to secure them. The four corners can now be put in from above to produce the position of Fig. 123. The middle layer is constructed by putting the edges on the corners just introduced—yellow-blue, green-yellow, green-white, and white-blue—thus completing the layer on top of the orange bottom (Fig. 124).

Four edges and four corners are now left to reassemble. To insert an edge cube on an incomplete face we repeat the dismantling move in reverse. We might begin with the white-red edge cube. We give the green face, now two-thirds restored, an eighth of a turn to make room for the edge cube (Fig. 125). We then complete the figure by inserting the white-green-red and red-white-blue corners. Next we give the yellow face an eighth of a turn to let us slip in the red-green edge and the red-yellow-green corner. Turning the blue face by an eighth, we introduce the red-yellow edge and the red-yellow-blue corner (Fig. 126). Finally, by turning the red face by an eighth and slightly pressing the red-blue edge into position, we restore the cube to its initial state (Fig. 127).

A first solution

We will now illustrate a simple sequence that transforms the initial state into a cross-pattern (Fig. 128), but first let us define our terms and symbols. Fig. 129 represents a counterclockwise quarter turn (90°) of the top layer. The circular dot in the center is the point of reference for the series of moves. A quarter turn of the side faces (right or left) is shown by an arrow as in Fig. 130. The move in Fig. 131 means a counterclockwise quarter turn of the front face, and the dotted arrow of Fig. 132

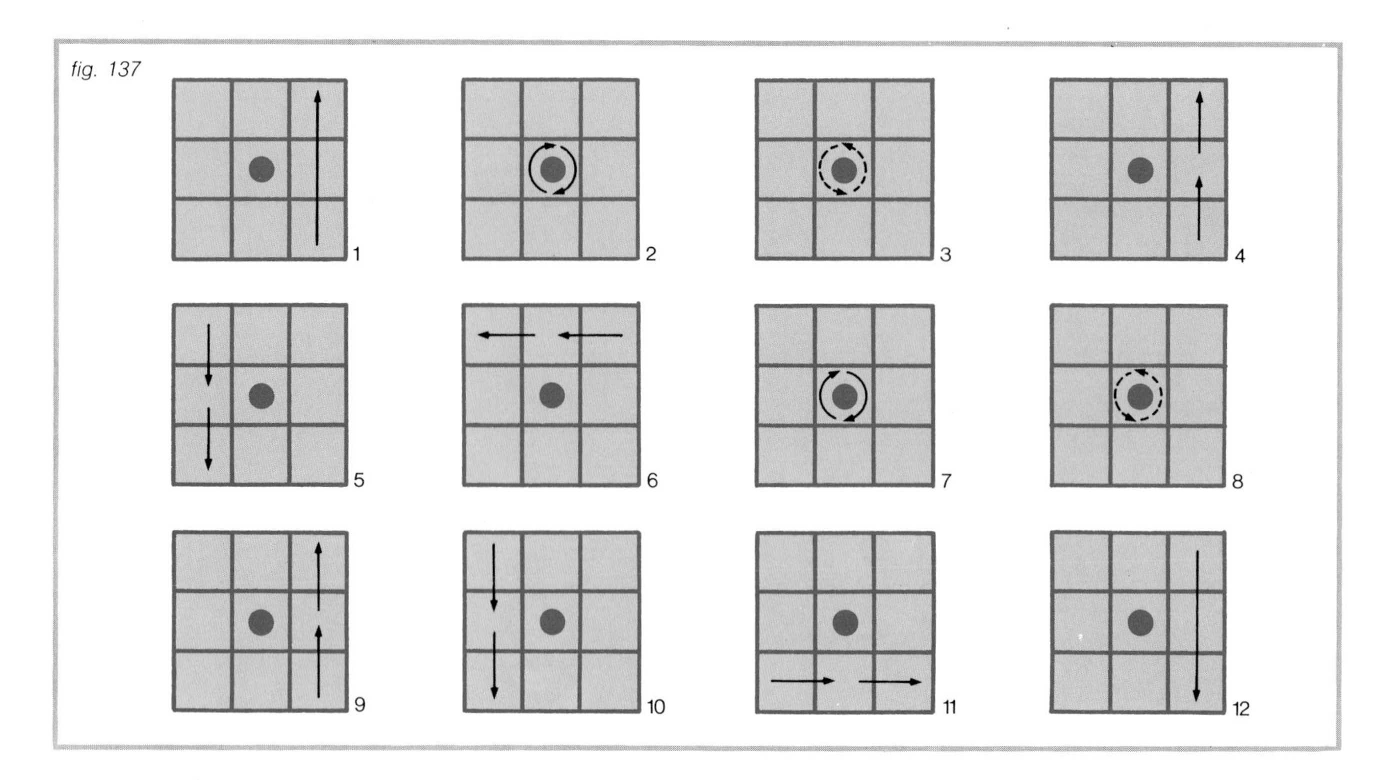

fig. 137

means the same for the back face. A rotation of the central layer is indicated as in Fig. 133. Finally, a half turn (180°) is shown by doubling the arrow as in Figs. 134–136.

The operation that must be executed has twelve moves. It begins with any face, which then remains the reference frame for all subsequent moves (Fig. 137). The result is the same if the sequence is performed in the reverse order—from 12 to 1. This is a first introduction and deceptively simple. If you experiment with moves, you might often feel anxious and bewildered. It is wise, therefore, to learn the movements of the 26 cubes and how to dismantle and reassemble the entire cube swiftly and smoothly. Then you can try simple problems, but do so with the help of a good instruction book. A wide range of literature is now available to guide the reader in his juggling with the more than 43×10^{18} arrangements.

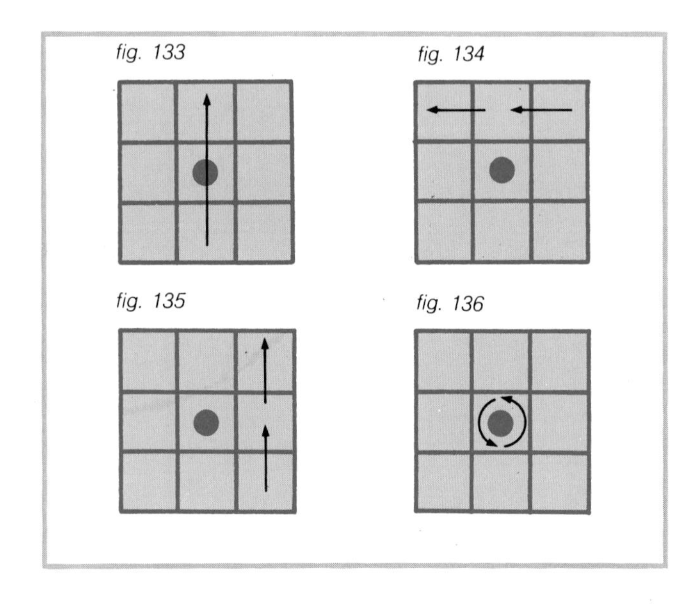
fig. 133
fig. 134
fig. 135
fig. 136

Finally, to really understand the cube, you must go beyond the established moves and consider the mathematical theories that explain its great variety of solutions.

PARADOXES AND ANTINOMIES

The infinite! No other problem has moved man's mind more deeply. (David Hilbert).

The role of paradoxes in the development of mathematical thought

In earlier chapters we dealt with games and mathematical questions that were clearly paradoxical, namely problems or results that contradicted the evidence or belied ordinary intuition and were therefore somewhat surprising. One cause of paradoxes in number games is the division by zero (cf. p. 28). It is a meaningless operation and, if used, can lead to any conclusion: $1 = 2$, for example. With figures, we saw the odd result from rearranging areas (pp. 51–52) where something seemed to spring from nothing; the error was in the incorrect geometrical assumptions we made.

The history of mathematical and geometric thought is replete with logical difficulties called paradoxes and antinomies. The word paradox comes from the Greek and means "beyond belief." It refers to assertions that contradict common sense or elementary principles of logic; antinomy, also Greek, is a legal term and means "against the law," though it is now a philosophic term for a logical contradiction between two equally valid assertions. Human thought tends to resolve such difficulties. Here it can be assumed the two terms mean almost the same.

Historically, paradoxes and antinomies have revealed countless hidden logical difficulties. Mathematicians and logicians have often had to reclarify problems and reestablish theories, at times producing a genuine renaissance not only in the field of mathematics but in all the sciences.

Here we shall look at a few paradoxes of ancient Greek mathematics and logic and one modern antinomy, discovered by Bertrand Russell in 1902, concerning some basic concepts of set theory.

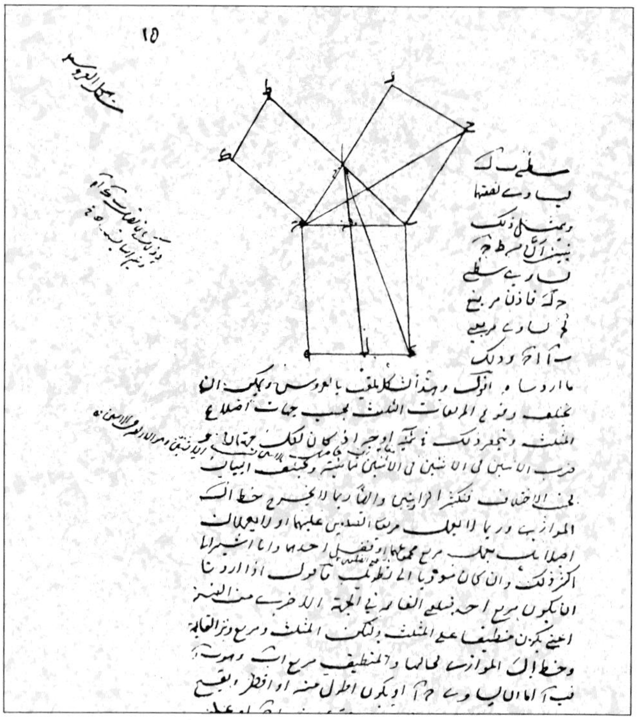

Left: Pythagoras' theorem as it appears in Ishaq Ibn Honan's 9th-century translation of Euclid's *Elements*.
Right: A visual representation of Pythagoras' theorem. We quickly see that the square erected on the hypotenuse of a right triangle is equal to the sum of the squares on the other two sides. The small squares directly link the figure with the numerical expression and thus confirm it.

Pythagoras and Pythagoreanism

An early and famous mathematical paradox was revealed by the Pythagoreans, the founders of Greek mathematics. To put the paradox into context we must look briefly at Pythagoras and the doctrines of his followers.

Pythagoras is noted for the theorem bearing his name. In fact, the theorem was known centuries earlier to the Babylonians under Hammurabi and also to the ancient Egyptians. Pythagoras was born in the mid-6th century B.C. on the island of Samos, off the west coast of Asia Minor; about 530 B.C. he journeyed to Croton in Magna Graecia (Southern Italy) to escape the tyrant Polycrates and his autocratic government. At Croton, Pythagoras founded a religious society whose members, chosen for their morality and intellectual dedication, were bound by strict rules and pledged to observe sacred silence and accept the founder's doctrine, as well as follow certain practical rules aimed at ascetic perfection and preparation for the beyond. The doctrine of metempsychosis (the transmigration of souls) is Pythagorean; it states that the soul does not die with the body but enters other human or animal organisms. This produces a certain form of justice, as the merits and flaws acquired in earthly life reveal themselves with reincarnation. While a meritorious soul migrates into another human form, one without merit lives in the guise of some unpleasant animal.

Discoveries made in Pythagoras' school, and the nature of the studies were kept secret, indicating the Pythagoreans were a sect. Indeed, legend tells of the disasters befalling those who dared reveal the society's insights.

The basic proposition of Pythagorean doctrine is that number is the essence of reality. Numbers are not only the inner principle of all things, but the very stuff of which they are made, namely the material points—infinitely small but not sizeless—comprising all bodies and elements. Numbers are classed as odd or even but any given number can be obtained as a composition of both odd and even. From this the Pythagoreans con-

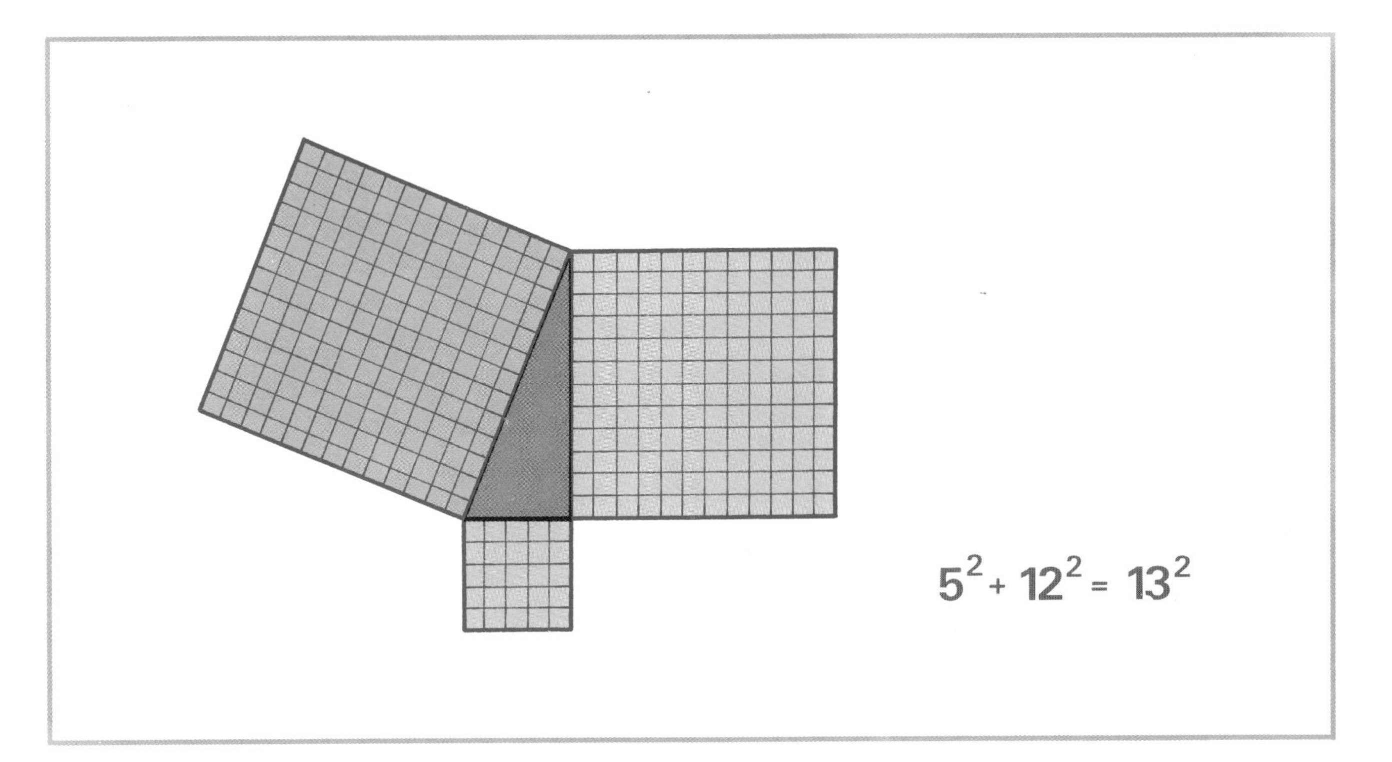

cluded that the odd and the even are the universal elements of numbers, and therefore of all things in the world. Moreover, they identified the odd with the finite, and the even with the infinite, as the odd sets a limit to division by two, while the even does not; the theory asserted that everything consists of finite and infinite. Similarly, they identified the odd with perfection and the even with imperfection. A later commentator explained it this way: The odd is perfect because it has a center of symmetry; among odd numbers, the most perfect is the triad which has a beginning, middle, and end. Even numbers, on the other hand, are imperfect because they can be divided into two equal parts, hence they lack proper structure.

Everything is comprised of opposite elements. The basic opposites which explain both objective and subjective reality, are contained in the sacred number ten:

1) finite and infinite,
2) odd and even,
3) single and plural,
4) right and left,
5) male and female,
6) rest and motion,
7) straight and curved,
8) light and dark,
9) good and bad,
10) square and oblong.

According to Philolaus, a 5th-century Pythagorean: "The decad is the foundation of all things . . . principle of divine, celestial and human life at once . . . without it everything is indeterminate, dark and closed. . . ." Indeed, 10 contains an equal number of odd and even; it has the unit and the first even number, the first odd number and the first square, thus making it the basis of all numbers. It is not surprising then that the mystic symbol of the Pythagoreans was the tetractys, representing 10 as the sum of the first four numbers.

If there are opposites, there must be a link to

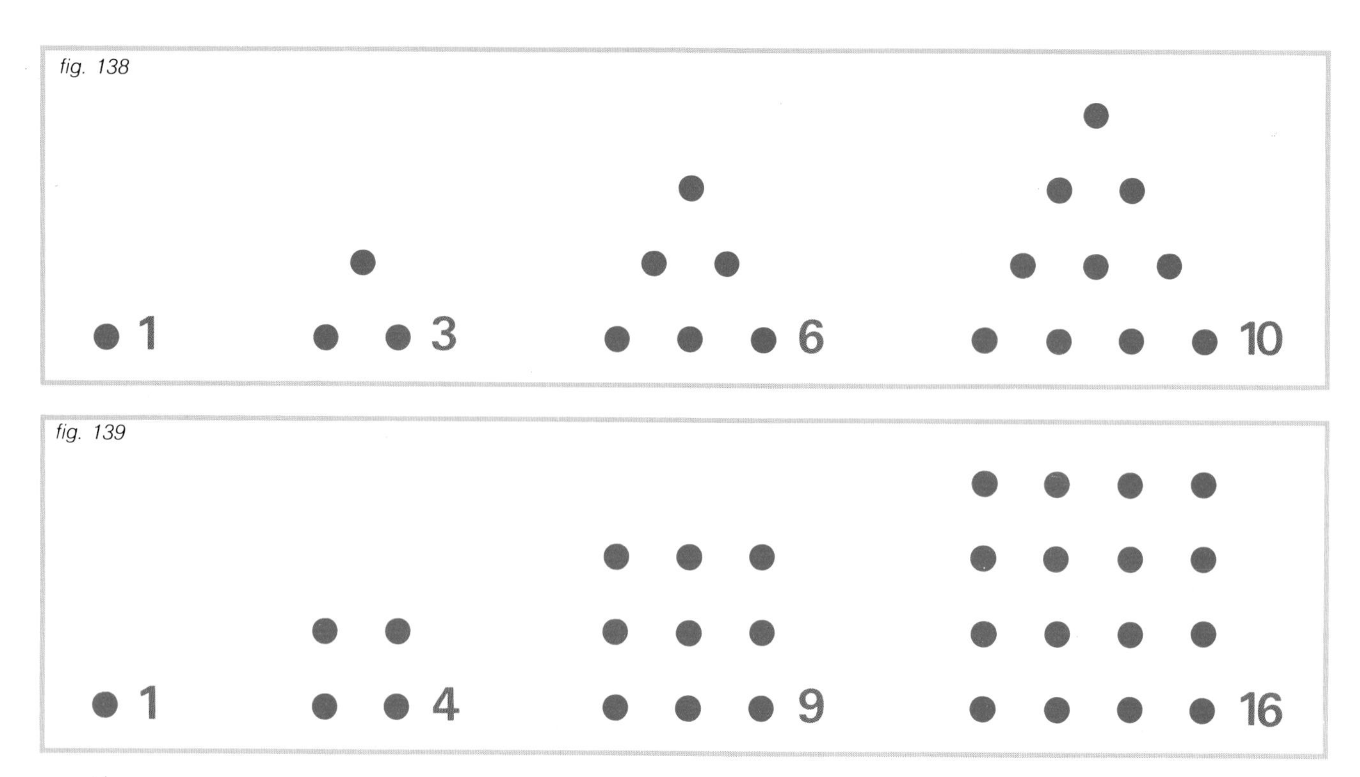

unite them. This is harmony, defined as "unity of the manifold and concord of the discordant." Thus, everything is number and harmony. Every number is a definite conjunction, a harmony of odd and even.

Geometrical representation of numbers

In later developments the Pythagoreans extended the theory of numbers to geometry, stating that geometric figures and relations between them are determined by numbers. Their triangular numbers suggest close bonds between arithmetic and geometry (Fig. 138), as do their square numbers (Fig. 139). Although these representations are much older than Pythagoras, and can be seen on pieces of Neolithic pottery, still, it was the Pythagoreans who illuminated their arithmetic features. The use of the abacus, which predated Pythagoras, is also linked to the geometric representation of numbers (cf. pp. 14–20). The formation of squares as the sum of successive odd numbers is probably derived from the abacus. Moreover this explains the importance of the gnomon (cf. the black lines in Fig. 140) as a generator of squares. "Gnomon" is Greek for "indicator" or "(carpenter's) ruler."

Numbers were thought to be so powerful that a cult actually developed around them. The discovery that music could be stated in terms of numerical relations helped to create a mystical attachment to mathematics. The Pythagoreans saw music as purifying the soul, and those who chose to live by Pythagorean principles were given ascetic preparation through music.

A tragic Pythagorean paradox: The odd equals the even

The Pythagorean system, although elegant, had genuine inelegancies. The Pythagoreans held that numbers could be interpreted as sets of points (which they pictured as pebbles), each with its proper position. Geometrical figures could then be regarded as finite sets of points, placed like so many grains one alongside the other. This is why the ratio of lengths could be expressed by

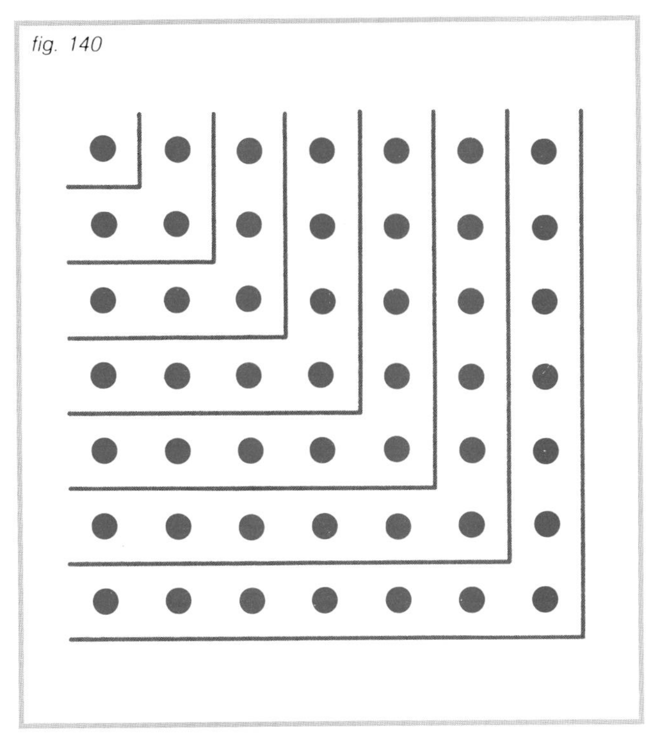
fig. 140

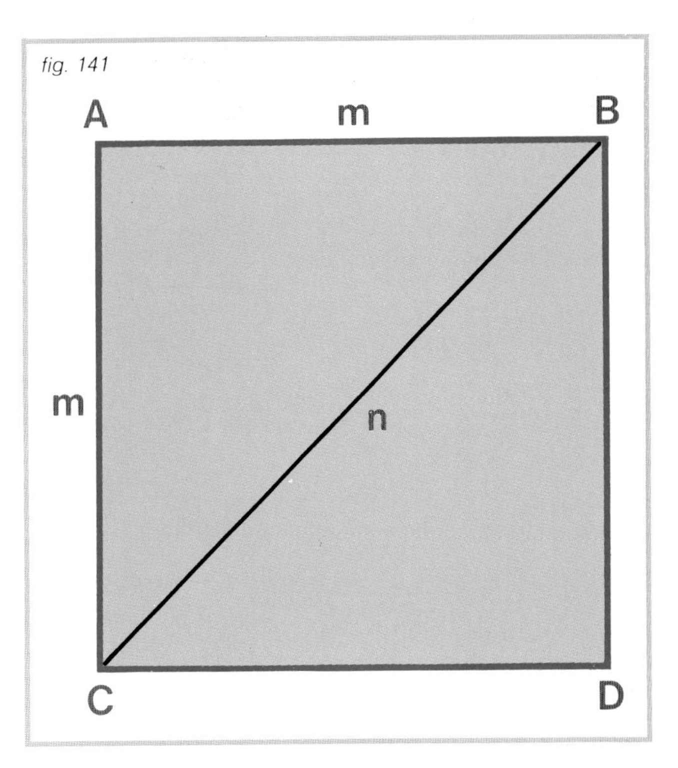

fig. 141

integers, the only numbers recognized by the Pythagoreans. For example consider two lengths, *A* and *B*. If they have a submultiple *C*, so that *C* is contained in *A* an integer number *a* of times and also contained in *B* an integer number *b* of times, the ratio *A*/*B* of the lengths is expressible as *a*/*b*.

Since the Pythagoreans believed geometric figures consisted of finite sets of points, they could not think of incommensurable lengths. They held that any two lengths must have a common submultiple, the "monad" or unit point of which all lengths are made up. When incommensurable lengths were discovered, it caused astonishment and upheaval in the school: lengths that could not be compared in terms of integers? It is ironic that the very theorem bearing the founder's name was the one to shake the foundations of "universal mathematics" and undermine the concept of a segment as a finite set of points and hence expressible by numbers.

Consider the square *ABCD* (Fig. 141) and take the side *AB* as consisting of a whole number, *m*, of points and the diagonal as consisting of *n* points. By Pythagoras' theorem, $2m^2 = n^2$, or, removing common factors in *m* and *n*, $2r^2 = s^2$, where *r* and *s* are relatively prime (without common factor). Now if $2r^2 = s^2$, then s^2 must be even making r^2 odd, so that *s* is even and if we write $s = 2t$, and substitute this for *s* we have $2r^2 = 4t^2$, or, simplifying it, $2t^2 = r^2$. By the same argument, r^2 must be even, which is absurd because we have just seen that it has to be odd. Hence the diagonal is not commensurable with the side of the square. This contradiction overturned Pythagorean mathematics and its claims to universal explanation. Here was the first example of a problem that could not be solved by numbers; it contained two real things, the diagonal and the side, the ratio of which could not be stated numerically. Since these two real things belonged to the very figures that were thought identical with numbers, the school itself was in ruins.

The sacred halo encircling numbers faded. According to tradition, the revelation of incommensurability by Hippasus of Metapontum was regarded as a crime. The school authorities, bent on maintaining the secret, asked Zeus, father of the gods, to punish the sacrilege of Hippasus.

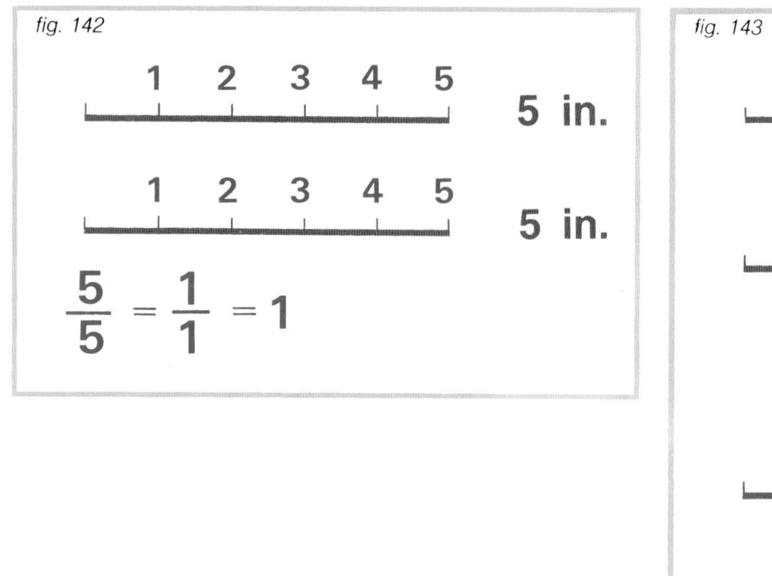

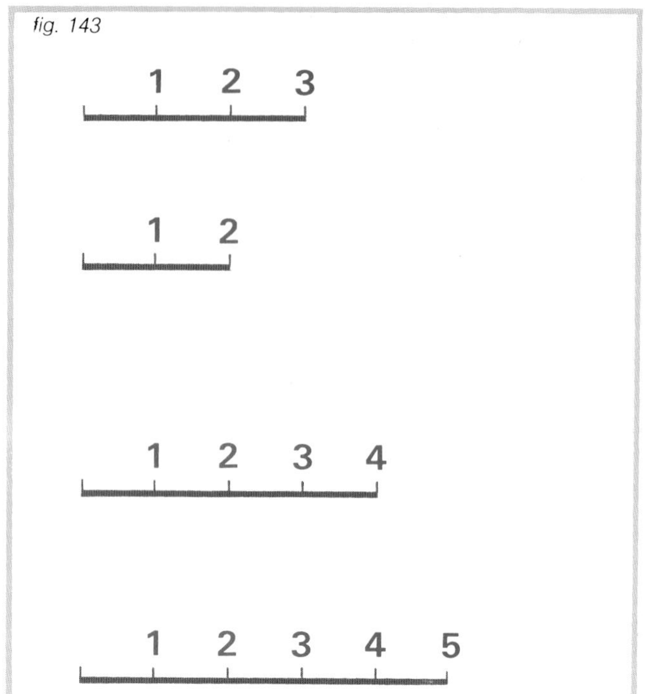

Zeus honoured the request and Hippasus was shipwrecked and drowned.

Unthinkable numbers

The Pythagoreans introduced the concept of the commensurable and applied it to the relation between figures. Less accurately we might speak of the measurable, that which can be expressed by numbers. This intuition incorporates the notion that reality can be described through the natural numbers. It must be remembered that for the Pythagoreans, mathematics was not merely a language but represented the very core of things; the mathematical description of an object was its essence. The scope of this intuition is hard to grasp, but we may reflect that the scientific and technical revolution was possible only because the most powerful branches of mathematics were applied to the external world. This occurred in the 17th century when Galileo Galilei (1564–1642) clearly saw that mathematics and geometry were the most effective and appropriate ways to grasp nature: "Philosophy," he wrote, "(meaning the natural sciences) is written in this greatest book constantly open before our eyes (I speak of the universe), but we cannot understand it unless we first learn to understand the language and to know the characters in which it is written. It is written in the language of mathematics and the characters are triangles, circles and other geometric figures . . . without them we vainly roam through a dark maze. . . ."

Describing objective reality with numbers rests on the basic concept of measurement or commensurability. For a segment, or anything, to be represented by numbers, we must assume it contains elements that can somehow be made to correspond to numbers. Hence the idea of the segment as composed of a finite set of points, such as grains. The granular notion of the straight line is based on this principle: Given a straight line, the finite number n of points in it can be taken to represent the line itself. The same is true for two segments; we can express the relation of their lengths with natural numbers. If the segments are equal (Fig. 142), then the two numbers are 1 and 1, for the two segments contain the same number of points; the ratio is 1/1. If the first segment contains the second segment twice or three times, the relation is expressed by the pairs (2, 1) and (3, 1), indicating that the first segment is twice or three times the length of the second. If the opposite is true, then the order of the numbers is reversed and the pairs are (1, 2) and (1, 3). In symbols, $a/b = 2/1; a/b = 3/1$; $b/a = 1/2$; $b/a = 1/3$. Until now we have considered segments, one of which was a multiple of the other. If this is not so, we halve one of them until we find a part exactly contained in the other. Take Fig. 143: The ratios are formed by (3, 2) and (4, 5) respectively. In the first case half of b is contained three times in a (or a third part of a twice in b). This should always be valid; we merely divide one segment into equal parts long enough to find a part, however small, that is contained exactly in the other. On this rests commensurability, the ability to find two integers to express the ratio. Measuring physical quantities follows the same procedure. If we say that a rectangular chart has a base of 175 inches we compare two segments, namely the one corresponding to the unit and the one corresponding to the side of the chart. The former (1 inch) is contained 175 times in the other. We use this kind of reasoning daily, but it conceals an entire philosophy of numbers and their role. What separates the Pythagorean concept from modern mathematics is the Pythagorean notion that numbers are real objects. There is a snare in this reasoning which the Pythagoreans themselves discovered. Indeed mutually incommensurable quantities whose ratio cannot be stated in integers do exist. An example is the side and the diagonal of a square. If we draw a square of side 1, the theorem of Pythagoras tells us that the diagonal is $\sqrt{2}$, which we know to be irrational (the Latin root of the word meant unthinkable), namely, not expressible as a quotient of two integers (Fig. 144).

The Pythagoreans held that "there is no part of the side contained exactly in the diagonal." We recall from school that to express $\sqrt{2}$ we must use numbers that represent it by excess or by defect: by excess 2, 1.5, 1.42..., by defect 1, 1.4, 1.41..., the former decreasing and the latter rising toward $\sqrt{2}$. If we reach a contradiction in the solution of a mathematical problem (in spite of the correct procedure), then one of the starting premises is false. In this case, the error was to assume that a segment is a finite set of points. It is at this juncture that we first meet the concept of the infinite.

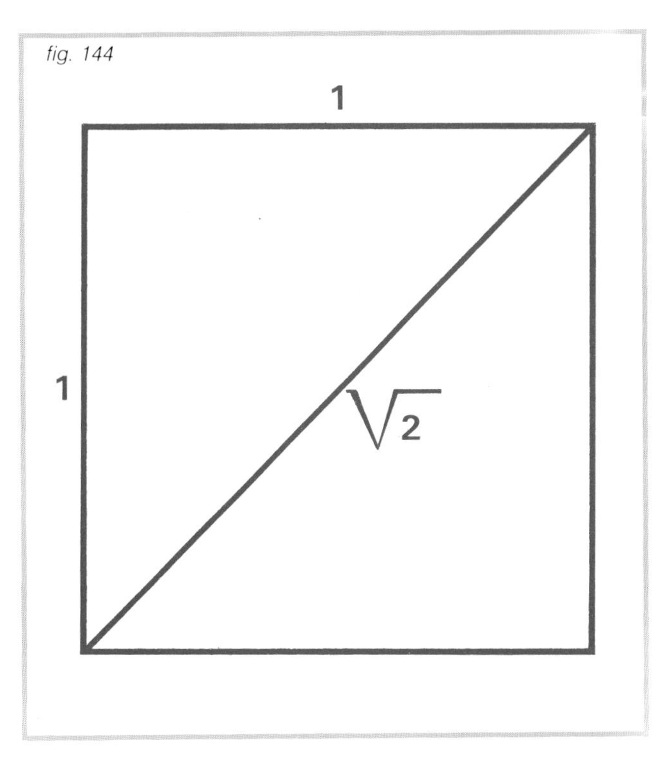

fig. 144

Zeno of Elea

The discovery of incommensurables presented Greek thinkers with new concepts of the infinite and the infinitely small or infinitesimal; witness the philosopher Anaxagoras who journeyed from Asia Minor to the Athens of Pericles after 450 B.C.: "As regards the small, there is always a smaller, and likewise for any large there is a larger." Such is the classical definition of infinity. These concepts were not developed any further by Greek mathematicians, but they formed the basis from which modern mathematics has grown.

Playing on the ambiguity of concepts such as

fig. 146

finite and infinite, Zeno of Elea (ca. 495–435 B.C.) introduced some paradoxes. The philosopher Parmenides, of whom Zeno was a faithful disciple, also lived and worked in Elea, in the region of Lucania. Constructing a doctrine opposed to the Pythagorean view of reality and phenomena, Parmenides stated that being, or rather the being of things, is one and indivisible. Among his followers, it was Zeno who ingeniously defended his master's doctrines against the Athenian philosophers.

Zeno's paradoxes

Roughly, a paradox involves a succession of steps in reasoning that contradict common sense and are therefore amazing or amusing. Zeno's paradoxes, however, apart from being amusingly presented, revealed serious difficulties ("aporiai" as they were then called) in Greek mathematics. The English philosopher-mathematician Bertrand Russell (1872–1970) commented on this in *The Principles of Mathematics*: "In this capricious world, nothing is more capricious than posthumous fame. One of the most notable victims of posterity's lack of judgment is the Eleatic Zeno. Having invented four arguments, all immeasurably subtle and profound, the grossness of subsequent philosophers pronounced him to be a mere ingenious juggler, and his arguments to be one and all sophisms. After two thousand years of continual refutation, these sophisms were reinstated, and made the foundation of a mathematical renaissance. . . . " Aristotle reports Zeno's paradoxes as: the dichotomy, Achilles and the tortoise, the arrow, and the stadium.

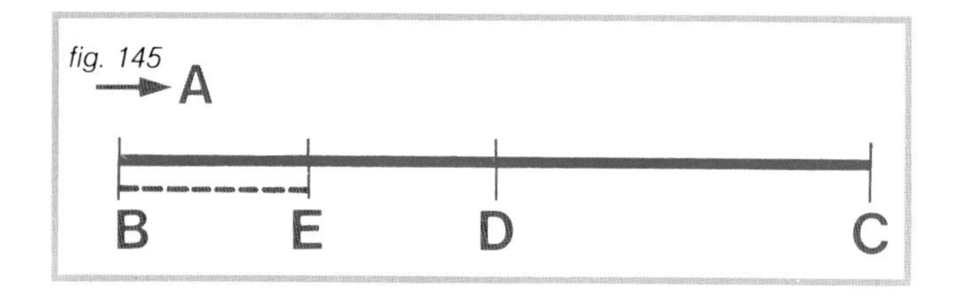

fig. 145

The paradox of dichotomy or bisection

We can formulate this as follows: Motion is impossible, for before reaching the goal one must reach

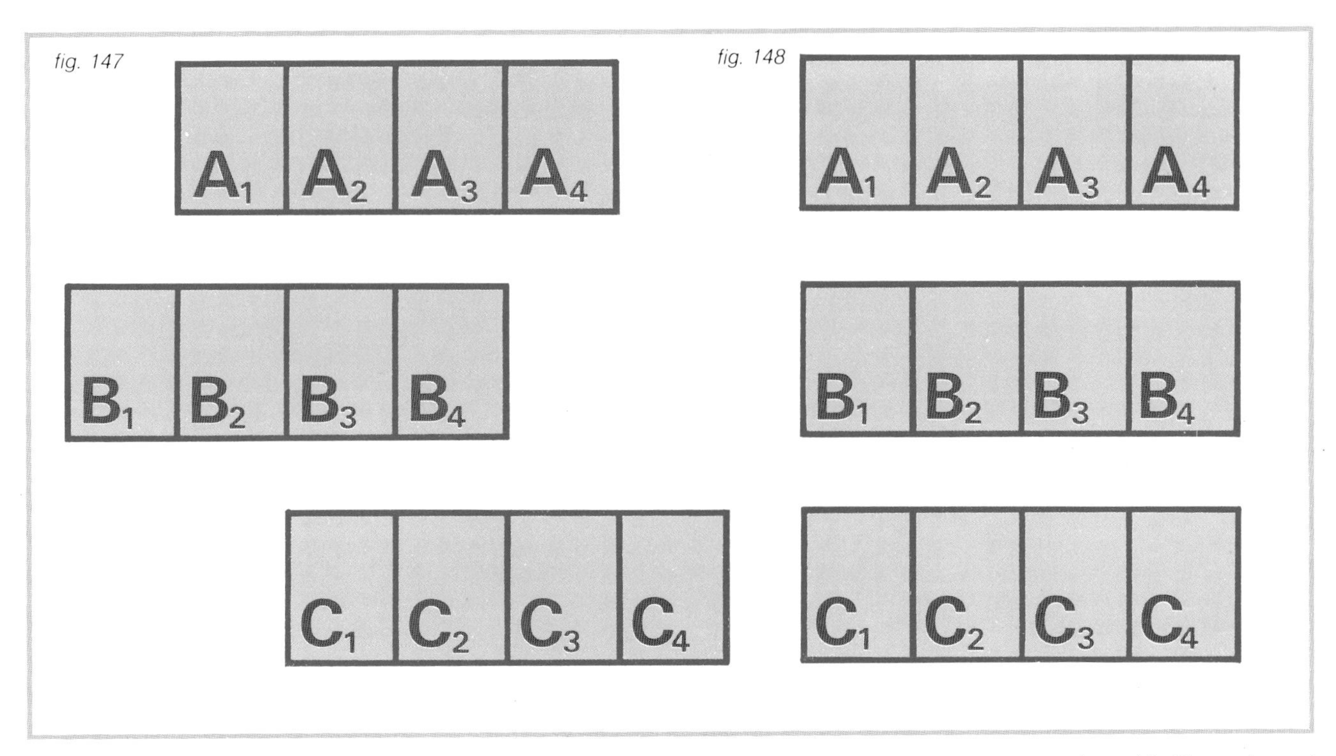

the halfway mark, and before that the quarter mark, and before that the eighth mark, and so on indefinitely. In Fig. 145, A is a moving point; it can never move from B to C, for it must first arrive at D, and before that at E, and so on. Suppose a runner must cover, in a finite time, the distance BC infinitely divided. This is clearly absurd, for no one can cover infinitely many elements in a finite time. Thus motion is impossible, even if our common sense tells us this is not so.

Swift-footed Achilles and the tortoise

The second paradox is the most famous. In a race between Achilles and a tortoise, Achilles will never catch up with the tortoise if the tortoise's starting point is ahead of Achilles' starting point. When Achilles reaches the tortoise's starting point, the animal will have gained a new starting point, a simple duplication of the initial situation. And so it goes, indefinitely. Although the distance between Achilles and the tortoise becomes ever smaller, it will never reduce to zero. Concretely, if Achilles (A) and the tortoise (T) move along a straight track at uniform speeds, with T moving at one-tenth the speed of A, give T an initial start of s. When A has covered the distance s, T has gone $(1/10)s$; by the time A covers this distance, T will have gone $(1/100)s$; while A covers this, T advances by $(1/1{,}000)s$ and so on continually. Therefore, says Zeno, "the slower runner must always be a little ahead." Thus A will never reach T, always remaining slightly behind however small the gap (Fig. 146).

The paradox of the arrow

The third argument states that the arrow always occupies a definite space and, as such, is stationary at any given moment. To move, it would have to be within and outside of its space at the same time. However, a sum of states does not give us motion and therefore motion is impossible.

The paradox of the stadium

The fourth paradox is the most difficult to state. This is a simplified version. Let A_1, A_2, A_3, A_4 be

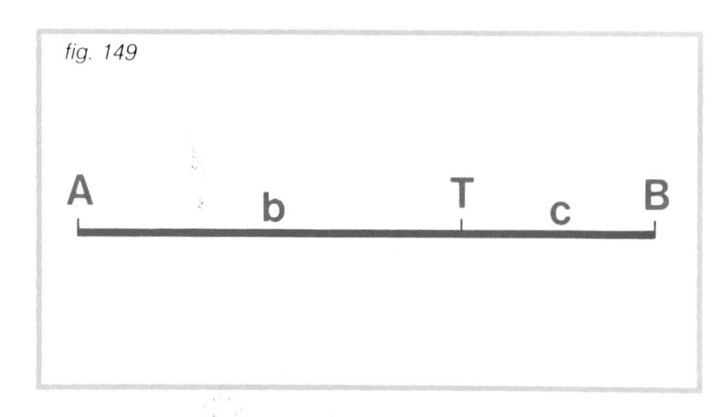
fig. 149

stationary bodies of equal dimensions. Let B_1, B_2, B_3, B_4 be of the same dimensions but moving to the right so that each *B* passes one *A* in the least time interval existing. Similarly, let C_1, C_2, C_3, C_4 be identical to the *A*s and *B*s but moving to the left at the same speed the *B*s are moving to the right. At a given moment, let the situation be as it is in Fig. 147. One instant later the position of Fig. 148 is reached. The *C*s and *B*s have moved in the same instant and now occupy the same position, but in this time lapse C_1 has passed two of the *B*s. Hence there is an even smaller instant, namely the instant it takes C_1 to pass one of the *B*s; and so it continues.

Theoretical significance and solution of Zeno's paradoxes

Certainly Zeno was not opposed to the concept of motion. Rather his paradoxes were aimed at the Pythagoreans who saw space and time as consisting of points and instants respectively. However, space and time are also continuous. The paradoxes merely make the concepts of finite and infinite, and indivisible and infinitely divisible as they were set forth in Pythagorean philosophy look ridiculous. In this sense Zeno's method anticipated the maieutic or dialectic method of Socrates who refuted his opponents by starting with their own premises and reducing them to absurdities. That these paradoxes were clearly aimed at Pythagoreanism is confirmed by their link to the concept of incommensurable quantities (p. 96), and to the attempts, (which proved unsuc-

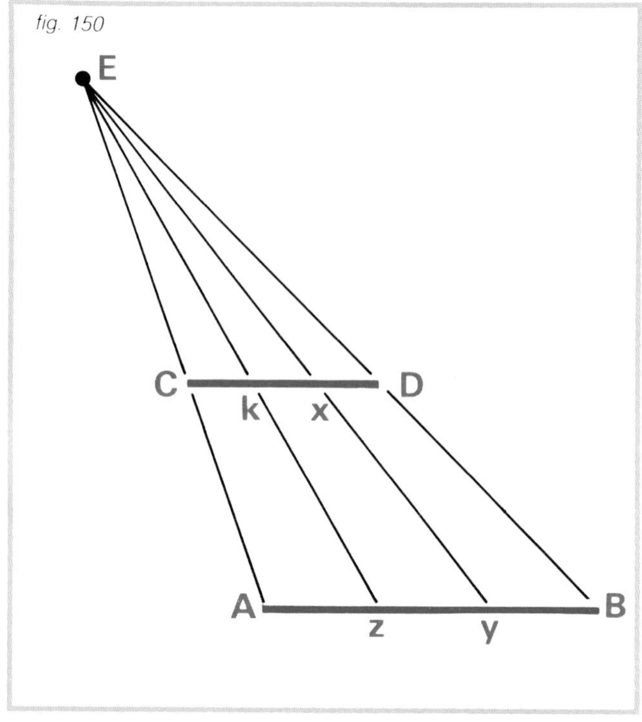
fig. 150

cessful), to eliminate such quantities by subdividing the unit indefinitely. The puzzling question of incommensurability led Aristotle to observe: "It seems indeed amazing that there should be something of which we cannot find the measure, however small it might be."

Zeno's paradoxes are based on precisely such logical difficulties: Indefinitely dividing a length never allows us to eliminate the residue as every division recreates the same problem in an infinite regress of the same logical difficulty, even though the quantity involved becomes increasingly small. Zeno's paradoxes, especially the dichotomy, and Achilles and the tortoise, are fully clarified by the modern theory of limits and its concepts of convergence and divergence of a sequence, which state that a convergent sequence has a finite limit, while a divergent sequence does not (it "tends to infinity"). There is an error at the root of Zeno's arguments which was elucidated only as mathematics developed, namely that an unlimited number of finite quantities adds up to infinity. In fact, if successive terms become ever smaller, the limit of their sum can be finite. Take the dichotomy and

fig. 151

1	4	9	16	25	36	49	64
↕	↕	↕	↕	↕	↕	↕	↕
1	2	3	4	5	6	7	8

fig 152

1	½	⅓	¼	⅕	⅙	⅐	⅛
↕	↕	↕	↕	↕	↕	↕	↕
1	2	3	4	5	6	7	8

suppose the whole track is of length 1. The successive stages are then, in sum: 1/2 + 1/4 + 1/8 + 1/16 + 1/32 + . . . , and this infinite sum adds up precisely to 1, the whole track.

Zeno's subtle arguments had a positive influence on future mathematicians. They were forced to stop using vague concepts full of logical traps, such as finite and infinite, indivisible and infinitely divisible; instead they had to develop clearer concepts and to adopt more rigorous methods.

The part equals the whole

Bertrand Russell, quite rightly, believed Zeno's arguments were immensely subtle and profound, particularly the well-known Achilles and the tortoise. The persistent gap created by the tortoise's advantageous starting point always repeats the problem. Here Zeno mocks the vain attempt to overcome incommensurability of two unequal quantities by reducing one to ever-decreasing fractions of the other. Experience and common sense may tell us that Achilles will catch up with the tortoise, but to assume this is not to deny the logical traps and paradoxical conclusions. Suppose, as the evidence teaches, Achilles reaches the tortoise at B when it has covered a stretch c (Fig. 149). Here we need a basic mathematical concept, to wit the intuitive concept of a set which we formally define below (p. 102). We all know more or less what this is. Let D be the set of points Achilles passes in reaching the tortoise's starting point and covering c, and let C be the set of points the tortoise passes in covering c. Then Achilles has covered $b + c$ while the tortoise has covered c. At any instant t, to any point in C there corresponds one and only one point in D. Hence c has as many points as $b + c$ which seems to contradict intuition; we imagine that the longer stretch has more points which is the same error the Pythagoreans made in assuming points have a small but finite size.

Mathematics forces us to abandon the intuition of physical things—a "point" is not a "thing"—that tells us the whole is greater than any of its parts. Galileo too reached this conclusion and offered an original demonstration of it: In simplified terms, take two segments AB and CD, with AB the longer. We might think of the respective distances run by Achilles and the tortoise. Connect A with C and B with D and extend them until they meet at a point E (Fig. 150). Take a point x on CD and let the extension of Ex cut AB at point y; similarly take z on AB and let Az cut CD at k. Proceeding in this fashion, every point in CD corresponds to one and only one point in AB and conversely: the part has as many points as the whole. Galileo transferred this conclusion to a comparison of natural numbers with their squares: there are as many natural numbers as there are squares, although the latter are included in the former (Fig. 151). Similarly the natural numbers, n, correspond to their inverses, $1/n$. As n becomes increasingly large though never infinite, so $1/n$ becomes increasingly small, although it never reduces to zero; indeed, zero is the limit to which the series of inverses tends. This "one-one" correspondence points out that there are as many inverses between 0 and 1 as there are natural numbers. Thus, even if intuition balks at the notion that a part has as many elements as its whole, the arguments are sound.

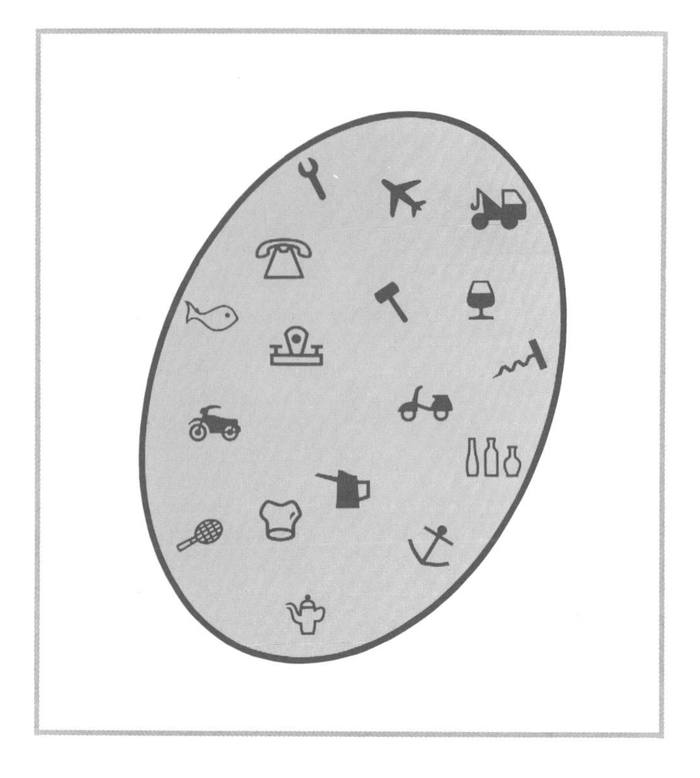

Sets: an antinomic concept

Finding an unambiguous definition for a concept has always been a fundamental problem. The basic concept of the new theory of sets was a source of difficulty for mathematicians and logicians in the late 19th century. We have used the concept above without defining it strictly; we must now do so in order to resolve the conundrum about part and whole.

What is a set? We use the concept daily, as have mathematicians and logicians from time immemorial. However, mathematics constantly widens its generalizations, and that requires strict definitions of the principles on which the development rests. From the start, the concept of set was fraught with logical traps and ambiguities. Even the ancients felt it; witness the sophism of the heap with which the sophists of 5th-century Athens amused themselves: "A grain is not a heap, neither are two, three, four, five grains, and so on. . . . Yet the heap is made up of grains." This plays with the fact that "heap" and "more" are imprecise concepts semantically linked with the concept "set." Mathematicians avoided the task of defining the concept "set" until Georg Cantor (1845–1918), in articles published from 1895 to 1897, attempted a general definition acceptable to all: *By a 'set' we are to understand any collection into a whole M of definite and separate objects m of our intuition or our thought. These objects are called the 'elements' of M. In signs we express this as:* $M = \{m\}$.

At first, this appears to be an unexceptionable and transparent definition. However, there remains an ambiguity in concepts such as "collection," so the vexing problem of defining the concept "set" in a logically correct manner persists.

Many mathematicians refuse to deal with the obstacle and prefer to look upon set as simply a primitive concept with which the mind operates. However, by rejecting Cantor's definition in an attempt to avoid possible antinomies, they manage

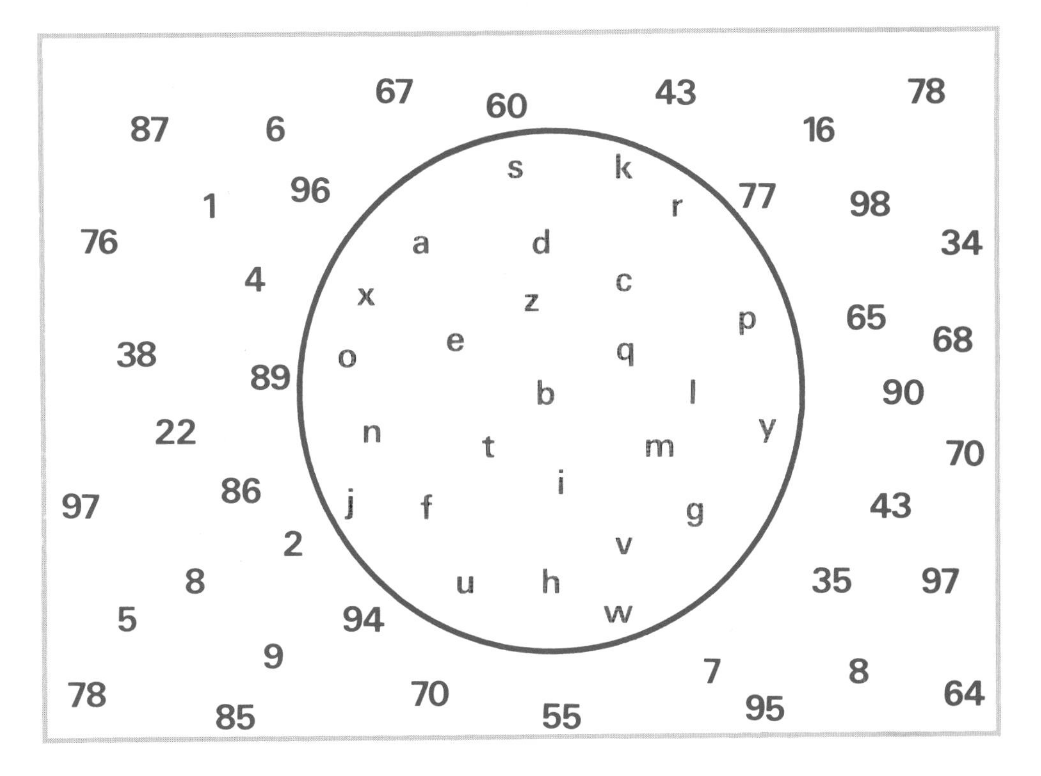

Sets can be intuitively represented with Euler-Venn diagrams (p. 114).
Left: Drawing of a set of arbitrary objects.
Right: The set of the letters of the alphabet, to which numbers do not belong.

to create contradictions among the principles and the laws of thought.

About 1900, Gottlob Frege, using Cantor's results, had presented in his forthcoming book, what he thought to be a systematic account of mathematics. However, just prior to publication Frege received a letter from Bertrand Russell reducing the author's construction to ruins. "To a writer in science," Frege admitted, "hardly anything is more unpleasant than having one of the fundamental elements of his construction shaken after the work is finished. That is the position in which I have been put by a letter from Mr. Bertrand Russell." The letter contained a famous antinomy now known as Russell's paradox. Let us first consider its popular version.

A postman and barber in trouble

In an imaginary village, there are two strange rules: The only postman employed by the town council must deliver the mail to those who do not call for it at the Post Office; the only local barber must shave those, and only those, who do not shave themselves. One day the barber said to the postman: "I am in a muddle and I don't know how to get out of it. The council says I am to shave only those who do not shave themselves. What about me? If I shave myself, I belong to the self-shavers and therefore I am not allowed to shave myself. If I do not shave myself, I am one of the others, and therefore I must shave myself. What am I to do?" The postman replied: "I am in a similar fix. I must deliver the mail only to those who do not call for their mail. What about letters addressed to me? If I get them at the Post Office, I am one of those who call for them and therefore I cannot deliver them to myself. If I do not call for them, I belong to the other lot and therefore I must deliver them. May I, or may I not, take my own letters?"

The reasoning moves in a vicious circle of contradiction and no logical escape seems to be possible: if yes, then no; if no, then yes. Let us look at Russell's abstract case.

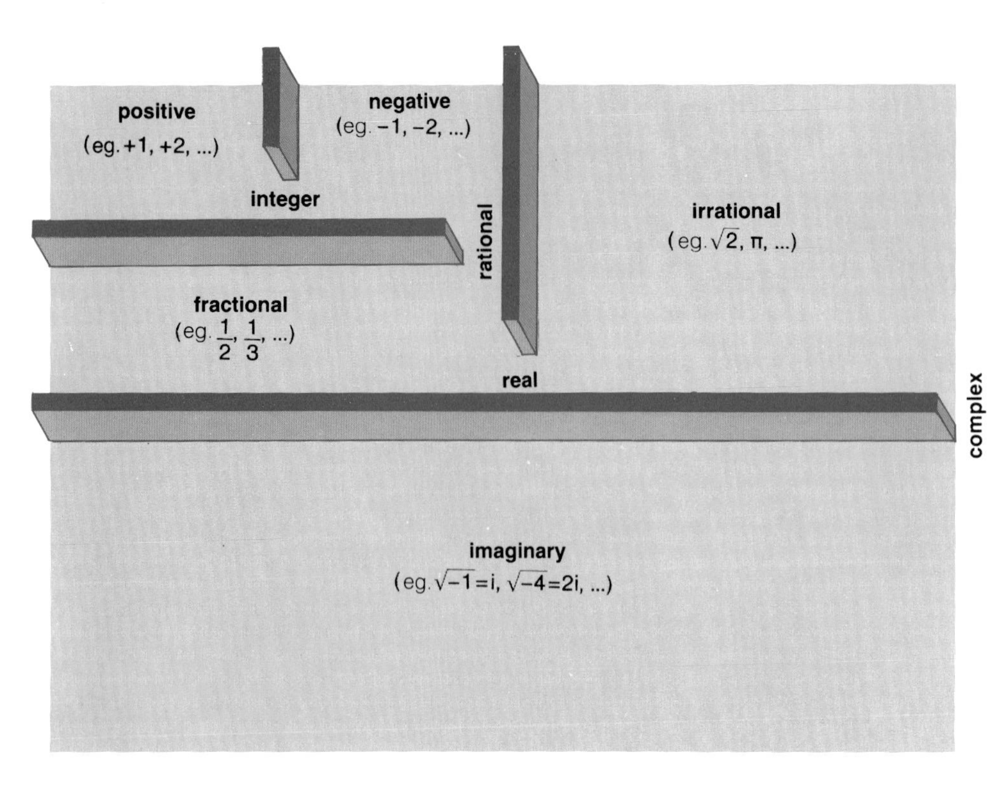

The system of numbers (cf. p. 21). Historically numbers did not evolve in a logical order. Man began by using natural numbers (N). Next came a set consisting of positive and negative integers and zero, denoted by Z. We then move to positive and negative fractions, which with Z, form the set of rational numbers Q. Irrationals are those numbers incommensurable with rationals, and together they form the set of real numbers R. By introducing $i = \sqrt{-1}$ we produce imaginary numbers, and finally complex numbers, which have a real and an imaginary part. These are denoted by C.

Russell's paradox

Using Cantor's definition of set (p. 102), we can frame the concept of a set consisting of all those sets that do not contain themselves as members. For example, the set of men is not *a* man, and the set of trees is not *a* tree. Therefore, some sets lack the features of each of their elements. Take a linguistic example. The word "monosyllabic" is not itself monosyllabic, as it has five syllables. However, the word "polysyllabic," also of five syllables, is polysyllabic. Similarly, there are sets containing themselves as members. For example, the set of abstract concepts is itself an abstract concept; or another example, "the set of all objects which can be described in thirteen English words," which itself can be described in thirteen English words. Sets not containing themselves as members are called "normal," while those containing themselves as members are "non-normal."

Consider now the set N consisting of all normal sets. We can ask whether the set N is itself normal or not. If N is normal then it belongs to the set N of all normal sets; but that means that N contains itself as a member, hence that it is non-normal; a contradiction! If, on the other hand, N is non-normal, it does contain itself as a member, and therefore belongs to N. But that means it is normal, and again we are faced with a contradiction. Such paradoxes are key in the history of mathematics, and in the early decades of the 20th century they shook the foundations on which mathematicians tried to build the entire logical and symbolic apparatus of mathematics. Later developments in this field testify to the fruitfulness of the crisis.

A GREAT GAME: MATHEMATICAL LOGIC

There exist, indeed, certain general principles founded in the very nature of language, by which the use of symbols, which are but the elements of scientific language, is determined. (George Boole)

A special chessboard

Consider the game of chess. It involves a board, two sets of pieces, and rules governing the moves; each move leads to a new position and eventually to the end, with one player the winner and the other the loser (if not, the game is a draw).

Logic, the science of correct reasoning, can be compared to an immense chessboard. The rules are those of deduction and the pieces a set of premises given as true. The moves are those allowed by the rules and each move leads to a new proposition. For the game to yield any result, the rules must be precise and strictly observed. Each move takes us toward the desired conclusion.

What is a logical argument?

Consider the following arguments.

1) If I had wings like a seagull I could fly. Since I have no wings, I cannot fly.

However, I could take an airplane, hence, the argument is unsound.

2) If there is no electric current the train stops. The current is on, so the train will not stop.

This will not do either as the train can stop for reasons other than lack of current—when reaching a station, for example, or because of an obstruction on the track.

2A) If there is no electric current the train stops. The current has failed, therefore the train stops.

This time the argument holds. Electric current is necessary to run the train. If the current fails the train stops, although it might stop for other reasons as well.

3) If it snows it is cold. It snows, therefore it is cold.

3A) If it snows it is cold. It is not snowing, therefore it is not cold.

The first of these propositions is valid. Snow falling necessitates a temperature at freezing point or below, therefore it is cold. The second argument is invalid; it can be cold without snow falling.

Hence 2A) and 3) are valid (we shall give a formal logical definition of this below); logic enables us to state principles in a reasoned manner.

4) All Helen's friends are my friends. All my friends are boring, therefore all Helen's friends are boring.

4A) None of Helen's friends is my friend. None of her friends is boring, therefore none of my friends is boring.

The first is valid, the second is not.

All these arguments start from propositions that can be true or false. They are called premises. From these we deduce a final proposition or conclusion:

If there is no current, the train stops.
The electric current is on. } Premises

The train does not stop. Conclusion

While in 2A), 3) and 4) the conclusion logically follows from the premises, in 1), 2), 3A) and 4A) it does not.

Logic is an old science. The first person to study it systematically was the Greek philosopher Aristotle (384–322 B.C.). In his *Prior Analytics* he gave us the groundwork on which logic has rested for some two thousand years. Toward the mid-19th century with the general revival of science and technology that accompanied the Industrial Revolution, attempts were made to join traditional logic with the methods and developments of modern mathematics. From this there emerged mathematical logic, a field which has rapidly progressed ever since.

Logic and ordinary language

Everyday language is vague and imprecise. It must describe a host of objective and subjective situations and in doing so, several terms can be applied to a single concept, and several elements can represent a single mental operation. Take the following:

1) Snow is white.

2) John, fetch the milk.

3) The square of the hypotenuse of a right triangle is equal to the sum of the squares of the other two sides.

The first states a fact, as does the third: Of each we can ask if it is true or false, and we can then frame an answer based on the evidence of the senses and the principles of geometry respectively. The second is an order and the question of "true or false" is not applicable as there is no statement about objective reality.

Mathematical logic does not deal with all the propositions that can be formulated in ordinary language, but only the ones that can be judged true or false. This includes the propositions that make up science and, more generally, all the disciplines that present systematic knowledge of objective or subjective reality. Which of the following concern mathematical logic?

1) Did you go to the doctor?

2) A water molecule consists of two atoms of hydrogen and one of oxygen.

3) Go to hell!

4) Aeschylus wrote "Seven against Thebes."

5) Did Euripides write "Alcestis"?

6) $E = mc^2$.

Of these, 2), 4) and 6) are assertions and involve logic; 1) and 5) are questions and 3) is a command, hence, 1), 5) and 3) cannot be considered true or false.

Mathematical logic requires precise and rigorous reasoning. Above all it must simplify and, without ambiguity, define the concepts it uses. Notice, for instance, the ambiguity of the term "always" in the following:

1) It always rains when I want to go to the mountains.

2) Whatever number you write down, you can always write down a larger number.

In 1), the word "always" is more descriptive of an emotional predicament, while in 2) it describes a logical feature of natural numbers. In logic, "always" is used in the latter sense.

An ingenious idea of Leibniz

A means of overcoming the vagueness of spoken language was suggested by the noted German philosopher and mathematician Gottfried Wilhelm von Leibniz (1646–1716). He held that, "spoken languages, though generally useful for discursive thought, are nevertheless subject to countless ambiguities of meaning and cannot offer the advantages of a calculus which reside particularly in the ability to discover errors of deduction deriving from the structure of words.... This wonderful advantage has until now been offered only by the symbols of arithmetic and algebra, where deduction consists simply in using symbols, and an error of calculation is the same as an error of thought."

Although in the history of ideas, Leibniz is remembered primarily as a philosopher and as the inventor of the infinitesimal calculus, his work also contributed to a decisive moment in the development of logic. Leibniz set out to construct a universal tool, the *Characteristica universalis,* to serve as an artificial scientific language with symbols to represent a variety of given meanings "biunivocally." "Biunivocally" means that each symbol corresponds to one and only one meaning, and conversely. Hence, through a set of symbols, we might arrive at a faithful expression of individual ideas. Leibniz tells us that this insight came to him at age twenty: "One might invent an alphabet of human thoughts... and by combining its letters and analyzing the words so formed one could discover and examine everything." To understand Leibniz's enthusiasm, we must consider the historical setting. Geometry was being revised and systematized, particularly by René Descartes (1596–1650) who algebraized geometry. From this a technique called "analytic geometry" evolved. Using the same model, Leibniz tried to mathematicize human thought and thereby to extend to philosophy a method for obtaining all the sciences through a combinatory art, much as Descartes and others had done in arithmetic and geometry through algebra and analysis. Leibniz longed for a symbolic language which, with its rules of deduction, would let us analyze any form of reasoning with the same certainty we have in arithmetic and algebra. In short, the new mathematicized logic would permit one to proceed securely in arguments. "If a controversy arises," Leibniz wrote, "the discussion between two philosophers need be no more heated than that between two calculators. They only have to take up their quills, sit down before their abacuses and say: Let us calculate!" Thus a problem in philosophy becomes one in mathematics and is handled in a prescribed fashion. Leibniz never completed such a *Characteristica universalis;* indeed, he left no systematic account of it, merely traces and fragments. Nor was such a vehicle achieved by the next generation of mathematicians, although Augustus De Morgan (1805–1871) made important contributions. It was only with George Boole (p. 113), working 160 years after Leibniz, that the idea finally took concrete shape. When it did, mathematical logic did not prove to be the powerful tool for discovery and examination that Leibniz had so confidently predicted. Still, it showed that at least some human thought could be mathematicized, and on this basis mathematical logic was founded and developed.

Logic: the science of correct reasoning

Again, consider the following:

1A) If I had wings like a seagull I could fly. I do not fly. Therefore I do not have wings like a seagull.

2B) If the electric current is off the train stops. The train does not stop. Therefore the current is not off.

Intuitively we feel these are correct and later we shall formally prove that they are. In the chapter on games with numbers, we saw that by replacing numbers with letters the mathematical argument could be elevated to a more abstract and general level. The same is true in logic if propositions are represented by letters.

If A represents "Today is Sunday," and B "I shall go to the stadium," the compound proposition "If today is Sunday, then I shall go to the stadium" is: "If A, then B". The "If. . . , then . . ." is a logical connective usually represented by an arrow: $\rightarrow$. Thus we have $A \rightarrow B$, an operation called "implication." It means that if A is true then B is true: A implies B.

To indicate the opposite of a proposition we put the negation sign $\neg$ in front of it. For example: $\neg A$ or non-A, means "Today is not Sunday."

Next take the argument: If today is Sunday, then I shall go to the stadium. Today is Sunday. Therefore I shall go to the stadium. In symbols $A \rightarrow B$, A, $\therefore B$. (As in mathematics, the three dots mean "therefore.")

Going back to some of the examples above, if P = I have wings, Q = I can fly, R = The current is off, S = The train stops, the arguments are stated as they appear in Fig. 153. What does it mean to reason correctly? Does it depend on the content of propositions, or on their formal arrangement? In logic we distinguish between the correctness of propositions and the validity of an argument. A schema of argument is valid if it moves from true premises to true conclusions without exception. Should true premises lead to even one false conclusion however, the schema is invalid.

Opposite: How to take off a vest without taking off one's jacket. This exercise actually makes an amusing party trick. Topologically, it means that the vest was never "under" the jacket.

fig. 153

$$\frac{P \rightarrow Q \quad \neg P}{\neg Q} \qquad \frac{P \rightarrow Q \quad \neg Q}{\neg P}$$

$$\frac{R \rightarrow S \quad \neg R}{\neg S} \qquad \frac{R \rightarrow S \quad \neg S}{\neg R}$$

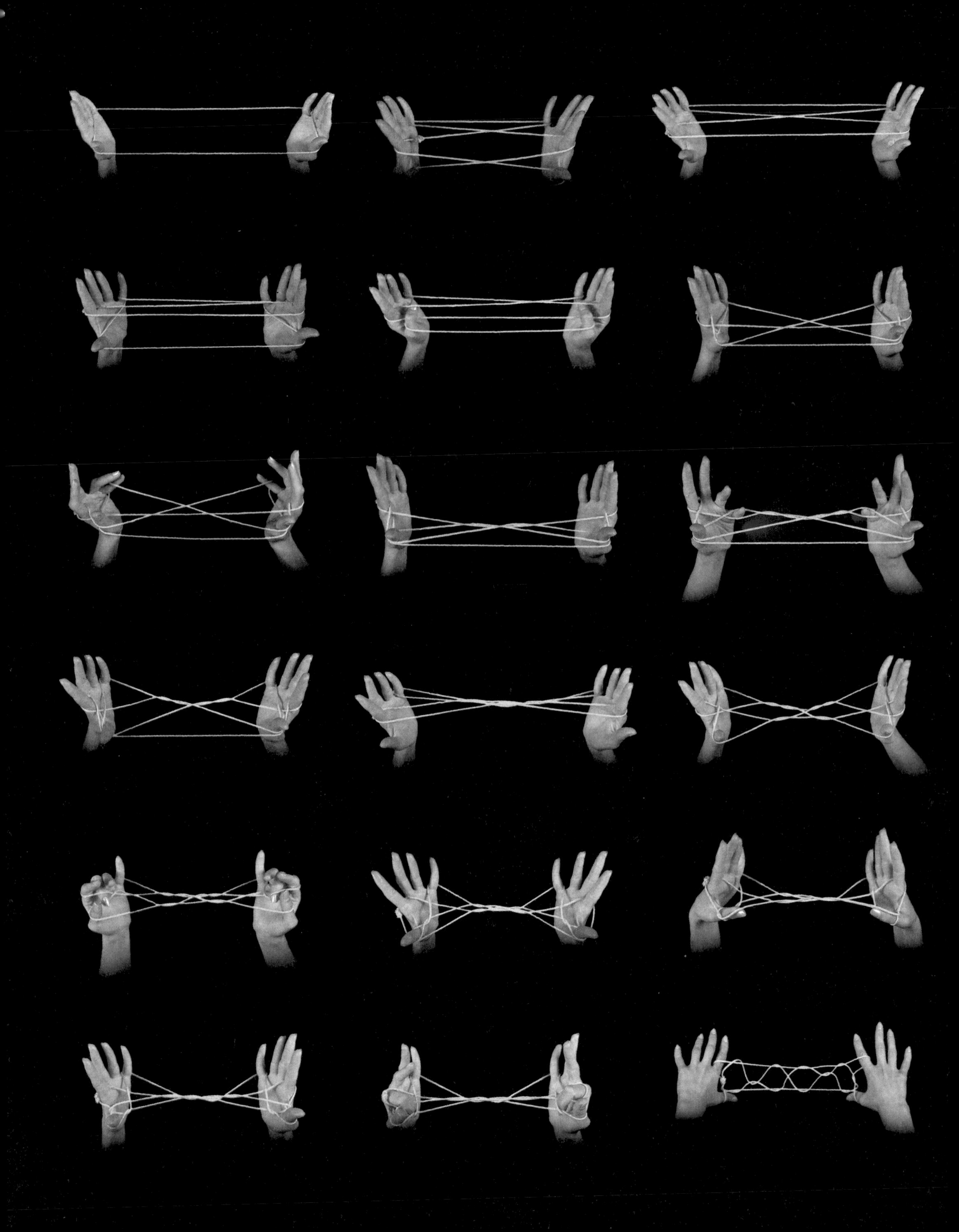

Left: Cat's cradles grew from the theory of knots and are a part of topology.
From left-to-right, top-to-bottom: A sequence leading to Jacob's ladder. 1) Pass the string behind the thumbs and little fingers. 2) With the index fingers, pick up the string crossing the opposite palm. 3) Let go of the string behind the thumbs. 4) Move the thumbs to the little fingers and with the back of the thumbs pick up the string between the little fingers. 5–6) Place the thumbs over the first string and pick up the second string from below. 7–8) Free the little fingers. 9) Place the little fingers over the last string. 10) Pick up the next to last string. 11) The thumbs are now free. 12) Pass the thumbs over the first two strings and pick up a third. 13) With the middle fingers, pick up the first string on the index fingers. 14) Carry this string onto the thumbs. 15) Turn the thumbs down. 16) Loosen a length of string from the thumbs. 17) Insert the index fingers into the triangle near the thumbs. 18) Free the little fingers and quickly turn hands down and forward. Jacob's ladder is now complete.
Right: Cat's cradles for two. Take the string (1) and make a double turn alternately with the right and left hands (2); with the back of the middle fingers pick up the thread on the opposite palm (3) and stretch it taut (4). The lower part of the figure should look like a roof. The second player begins (5) and inserts his thumbs and index fingers (6). The figure now resembles a window. With the left little finger the first player picks up the inner right string (7) and with the right little finger picks up the string on the left, and moving out tries to open the figure (8). The second player moves again (9) and with his little fingers seizes the upper strings and turning inward, further opens the figure which now resembles a butterfly (10). Using his thumbs from below (11), the first player then stretches the string (12). The second player, using his little fingers, takes the edges of the figure, while, from above the figure, he uses his thumbs and index fingers (13) to stretch the whole figure (14), which now looks like a hen's claw. Freeing the little fingers (15) the game ends.

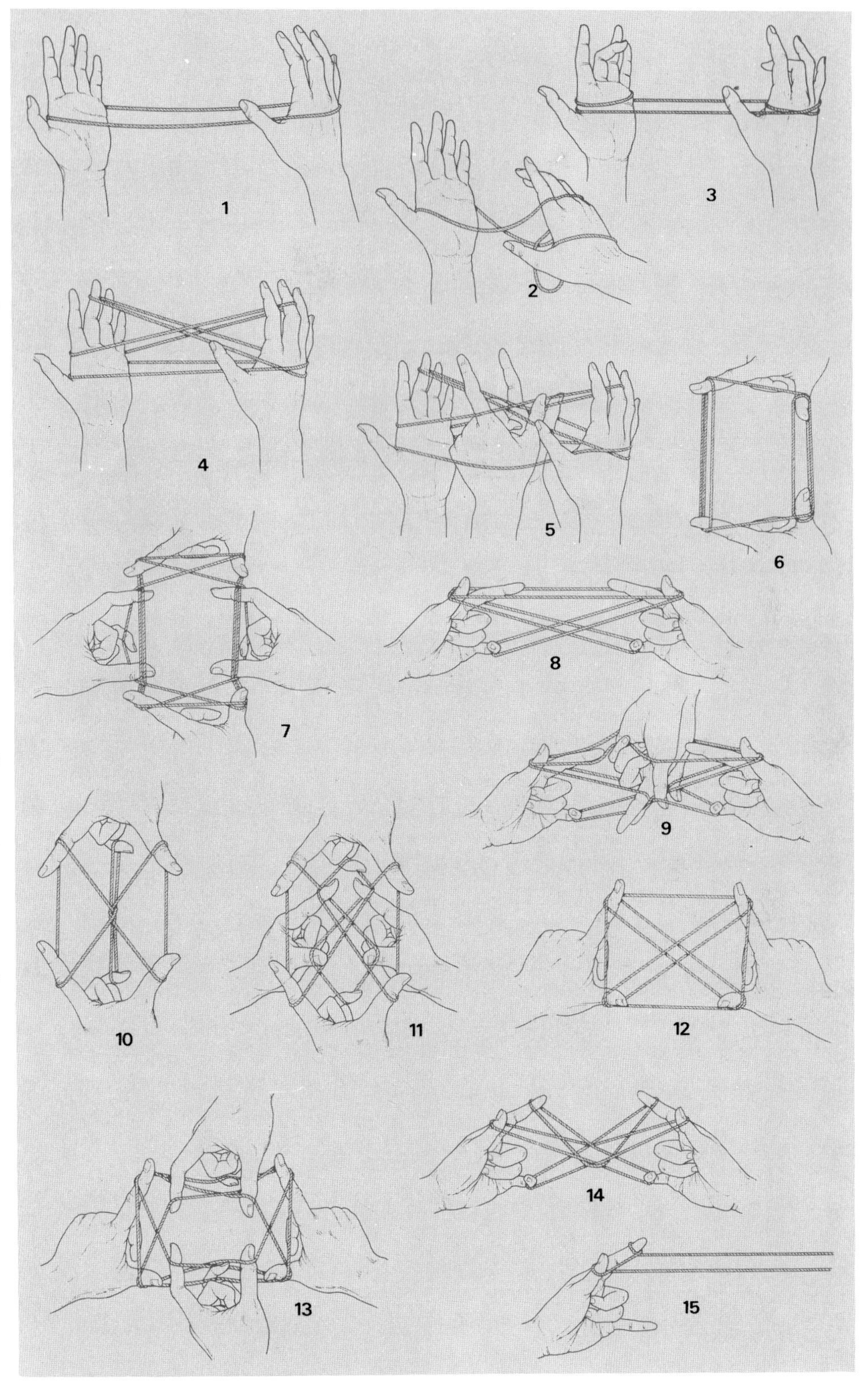

Mathematical logic simplifies and formalizes scientific language, which is why logic is good preparation for systematic study. Scientific theories let us forecast future events concerning the phenomena being studied. If the predictions prove accurate, the theory is confirmed; if they do not, the theory must be modified or discarded. Hence the importance of the rules and principles underlying the predictions. Mathematical logic, as the science of correct reasoning, has objective rules for reaching unquestionable conclusions. For example, if we know what forces act on a body at a given place and time and the velocity at which the body is moving, with a few mathematical terms and theorems and by applying logic to mechanics, we can predict the body's further motion.

Logical variables

We recall that a number can be represented by a letter—called a variable—if its value can take different numerical values. So too in logic. A proposition can be expressed by any letter which in this case is called a logical variable. Thus A, B, C . . . are logical variables that can be true or false; by convention, they assume the value 1 if true, and 0 if false. This representation facilitates economy and simplicity, and therefore generality. Without such symbols, reasoning would become very cumbersome.

Formal properties—irrespective of content or meaning—are the logical features that abstract schemata of reasoning share. The same is so in algebraic formulae, such as in $(x+y)(x-y)=x^2-y^2$, where x and y stand for any number.

Complex propositions can be constructed from simple ones through logical connectives linking two or more propositions, or operating on them. We have already met negation of a logical variable ($\neg A$) and implication ($A \rightarrow B$) which is read "If A, then B." If $A=1$, that is "if A is true," then $\neg A=0$, or "non-A is false."

Although we have given only the simplest and most elementary examples of arguments, logic can examine long and complex arguments. Indeed, a whole set of premises can be strung together, as in the following:

If John takes a cat home, he will neglect his homework. If he does not do his homework, he will get bad marks at school. If he gets bad marks at school, he will not move up. If he does not move up, he will not have a vacation. Conclusion: If John takes a cat home, he will not have a vacation.

This is illustrated in symbols in Fig. 154, with G = John takes a cat home, H = John will neglect his homework, I = John will get bad marks at school, L = John will move up, M = John will have a vacation.

The logical steps of this argument can move smoothly from one implication to the next because the operation of implication enjoys the so-called transitive property. (From $A \rightarrow B$ and $B \rightarrow C$, we can conclude that $A \rightarrow C$.)

Consider a similar argument in algebra, where we often draw conclusions that only mathematical logic makes explicit. Here the propositions are

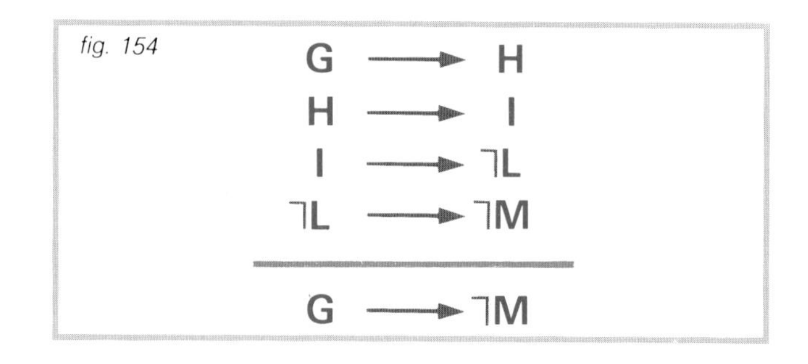

algebraic expressions, while the compound propositions to be proven are known as theorems. Given $2x + 9 = 13$, show that $x = 2$. Let D stand for $2x + 9 = 13$, E for $2x = 4$, F for $x = 2$. Now, algebraically: if $2x + 9 = 13$, then $2x = 4$, if $2x = 4$, then $x = 2$, therefore if $2x + 9 = 13$, then $x = 2$; and logically: $D \rightarrow E$, $E \rightarrow F$, $\therefore D \rightarrow F$. The conclusion is true only if all the premises are true; a single false premise will render the conclusion false.

George Boole and the origins of propositional calculus

The formal system built from logical variables—which assume the value 1 if true, 0 if false—and connectives representing logical operations, is known as Boole's algebra after the English mathematician George Boole (1815–1864). Boole was self-taught and acquired a vast store of knowledge, including mastery of Greek, Latin, French, German and Italian. In 1847 he published a small volume, *The Mathematical Analysis of Logic,* the aim of which was to investigate the fundamental laws of the mental operations involved in reasoning, to express them in the symbolic language of a calculus, and on this to construct rules for a science of logic. In *An Investigation on the Laws of Thought, on which are founded the Mathematical Theories of Logic and Probabilities* (1854), he gave a logical interpretation to his algebraic system, showing that it could be adapted to "bivalent" propositions, ones that can only assume two values, 1 if true and 0 if false. The idea was developed by his successors.

Before Boole, the role of logic in science was unclear, as were the connections between philosophy, logic and mathematics. However, Boole perceived logic as a branch of mathematics rather than philosophy, and he had a marked influence on the course of logic. If philosophy is the science of the existence of things and the search for what is, then logic must be classified with mathematics, as mathematics, and particularly algebra, rests on a set of symbols that can be operated on; logic too, can be studied through an independent symbolic calculus. Moreover, Boole grasped the limitations of traditional logic which was then still subject to Aristotelian and medieval rules and schemata; indeed, he demonstrated how these could be overcome with new and rigorous methods and with an algebraic calculus analogous to the one used in mathematics. His concepts, especially those concerning the close ties between logic and mathematics, form the basis for much that formal logic has since discovered.

The logical calculus

What began as a project with Leibniz, became concrete with Boole. Let us now follow Boole's steps and clarify some of his ideas which thus far we have used only implicitly. Boole had a very clear picture of a logical calculus and its peculiar features. In any reasoning, mathematical or geometrical, physical or philosophical, we start with premises from which we reach a conclusion. Logical analysis translates the premises into symbols and then, using the rules of logic, deduces the more important conclusions implied. With this model of analysis we can then assess the validity of the conclusions. Hence, interest shifts from content to formal procedure.

Logic is not primarily concerned with whether premises are true, credible or reliable, but rather with what can be validly inferred from them. In short, it studies the argument, not the meaning of the premises. This is the "formal" character of a logical calculus; its symbols are not interpreted, because to do so would be to give them a meaning, and from that we could not abstract the formal structure or basic schema of reasoning. In *The Mathematical Analysis of Logic,* Boole writes: "They who are acquainted with the present state of the theory of Symbolical Algebra, are aware that the validity of the processes of analysis does not depend upon the interpretation of the symbols which are employed, but solely on the laws of their combination. Every system of interpretation which does not affect the truth of the relation supposed, is equally admissible, and it is thus that the same process may, under one scheme of interpretation, represent the solution of a question on the properties of numbers, under another, that of a geometrical problem, and under a third, that of a problem of dynamics or optics." Thus Boole arrived at a purely formal calculus, which need not refer to quantities, numbers or geometric magnitudes, because it is based on symbols and rules obeyed by operations without any specified meaning. The formal, or mathematical, logic founded by Boole was later developed by Bertrand Russell (1872–1970), Giuseppe Peano (1858–1932), and Friedrich Gottlob Frege (1848–1925). The calculus based on bivalent propositions is called propositional calculus.

We can give a graphic interpretation of propositional calculus with Euler-Venn diagrams, which are set-theoretical notations for logical operations and relations and are now commonly used to teach mathematics from primary school on. We discussed Euler and his contributions earlier (pp. 54–55). The English logician John Venn (1834–1923), was ordained at 25 but resigned his ministry in 1883 to concentrate on logic which he taught at Cambridge. His two major works are *The Logic of Chance* (1866), and *Symbolic Logic* (1881). His diagrams for syllogistic inference are called Euler-Venn diagrams, having been derived from a similar device used by Euler but perfected by Venn.

Negation

This is the simplest logical operation. Given a proposition, negation changes the truth-value of it. Thus, if M = Today is Saturday, then $\neg$M = Today is not Saturday. In Venn diagrams, the two values a logical variable can have are seen in Fig. 155. The circle represents all Saturdays (days for which M is true) while $\neg$M (days for which M is false or, equivalently, $\neg$M true) is represented by the area outside the circle.

Whenever we have a property defining a set, we must first fix the scope of our consideration, or the universe of discourse; roughly, everything that is under consideration in a given discussion. The universal set U then consists of all the elements of the corresponding universe of discourse. In our example, the universal set contains all the days in the year, represented by the whole diagram. For a mathematician, the universal set consists of all numbers; for a botanist, all terrestrial plants. The complement, $\overline{M}'$, is then defined as the set of elements not belonging to M′, that is U − M′ (Fig. 155b).

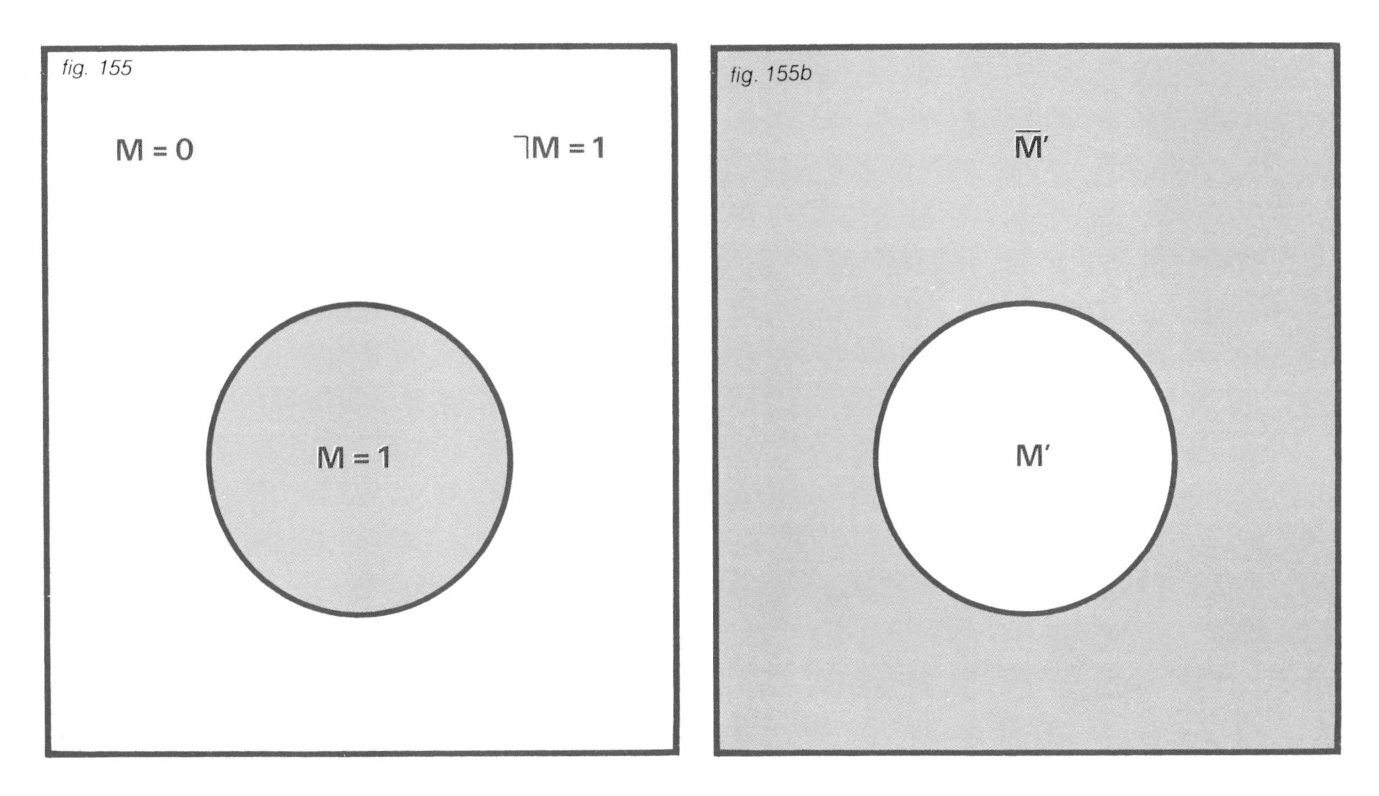

Explanation of symbols

It is advisable to use different symbols for different things. Since we have used capital letters for propositional variables, we will add an apostrophe when we wish to indicate the corresponding set. For example, if M = Today is Saturday, then M′ is the set of all Saturdays.

Conjunction and the empty set

Given two logical variables M and N, their conjunction is indicated by the connective $\wedge$, as $M \wedge N$. Thus if N = Today it is cold, then $M \wedge N$ = Today is Saturday and it is cold. In a Venn diagram the conjunction is represented by the intersection of the two sets, as illustrated in Fig. 156. M′ is the set of Saturdays and N′ the set of cold days. Then $M' \cap N'$ is the set that makes $M \wedge N$ true, namely all cold Saturdays.

The negation of $M \wedge N$ is $\neg(M \wedge N)$ and reads: It is not true that today is Saturday and it is cold.

Now try to show the following compound propositions graphically: John is staying in Australia and at the same time in England. Let E = John is staying in England, F = John is staying in Australia. Since you cannot be in two places at one time, the intersection of E′ and F′ is the empty set, $\varnothing$, which contains no elements and is a subset of all other sets (Fig. 157). An example from mathematics would be the set of numbers that are both positive and negative; of course, no number is.

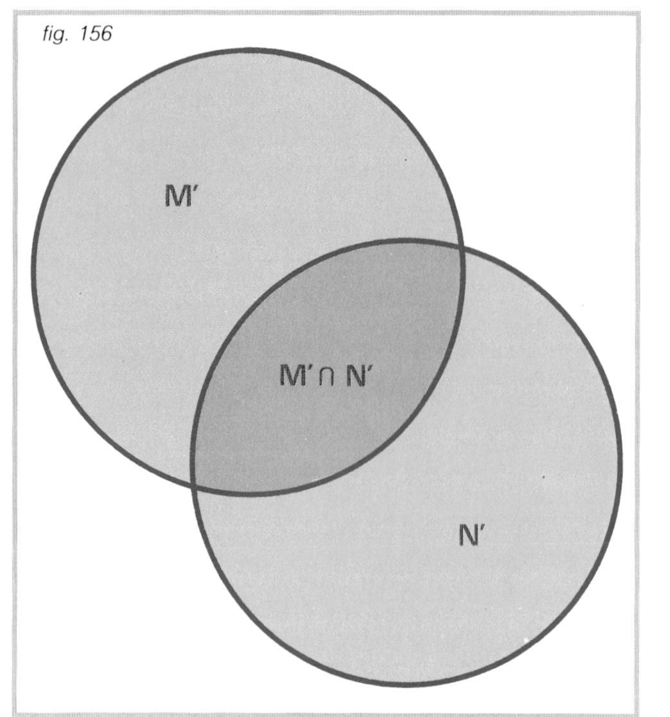

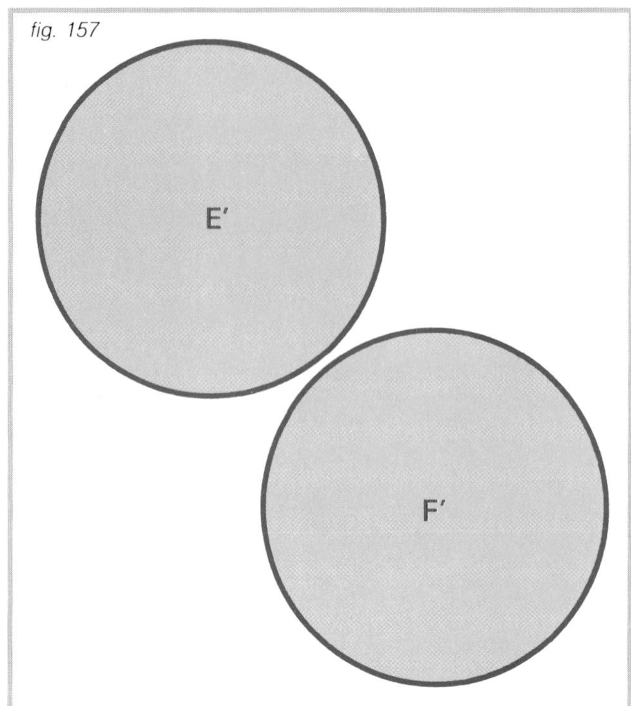

The empty set

The concept of the empty set leads us to the set-theoretical principle of extensionality. Suppose two sets are different. There must then be at least one object belonging to one set but not to the other; if two sets have the same elements they are the same set. Therefore, an empty set, which has no elements, is unique. We must not confuse the empty set ($\varnothing$) with the number of its elements (0). Like zero, the empty set is linked with nothingness—by definition that which does not exist—but nothingness is a concept and hence, we attribute an existence to it. In mathematics and philosophy as well as in everyday life, such concepts have proven eminently useful.

In set theory, it is the empty set that comes closest to nothingness. Much like zero for numbers, the empty set at least has the sort of being we ascribe to other sets, even if it is the only set without elements and a subset of all others, and thus somewhat different.

What does the empty set denote? In Fig. 156, M′ denotes the set of all Saturdays. But what does the empty set denote in Fig. 157? It simply "denotes" without referring to anything. As with zero, our reasoning becomes entangled and once again we are moving in the direction of paradox.

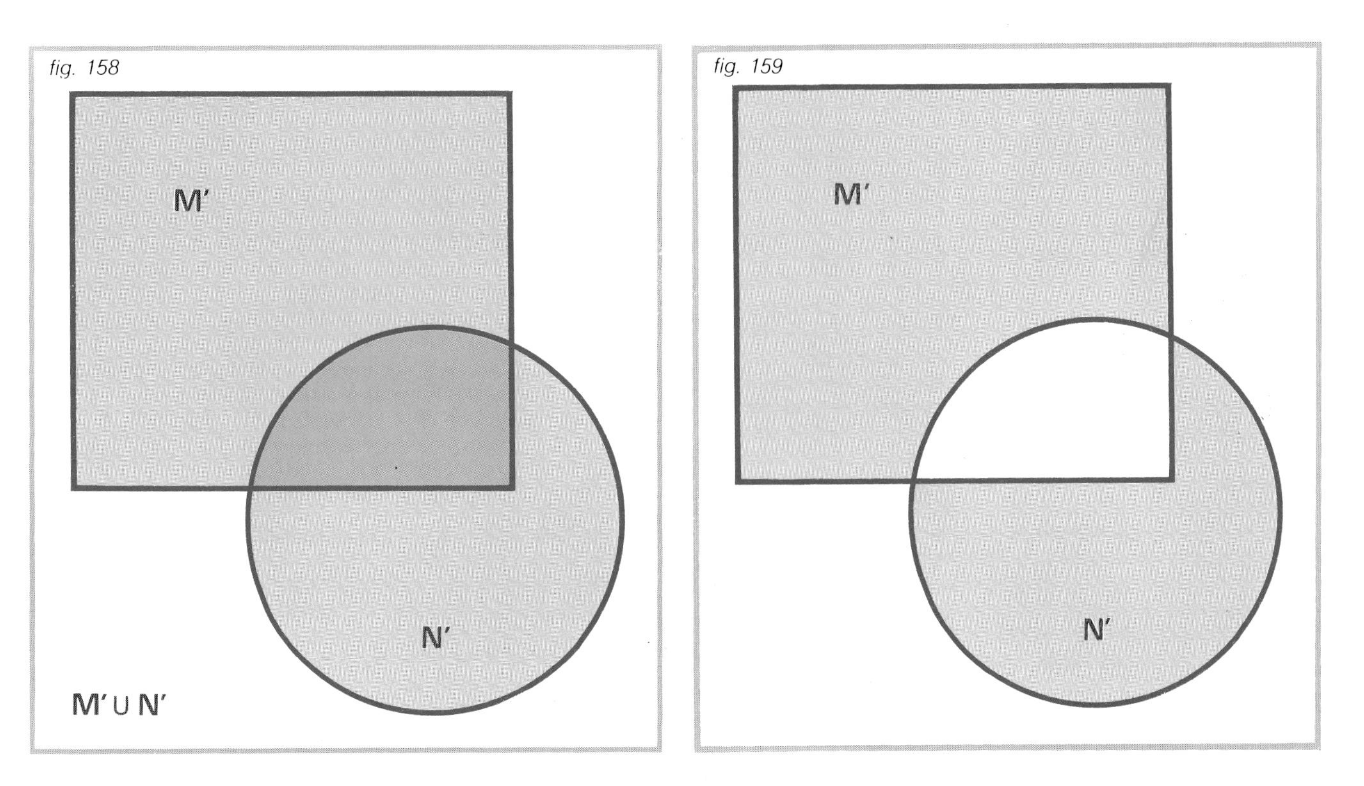

A first paradoxical consequence is that the set of dogs able to read this book is identical with the set of square circles, and each in turn is identical with the set of people living in England and Australia at the same time. To resolve the dilemma, we could say $\varnothing$ denotes the set of those elements that obey some contradictory assertion; or that whatever we assert about the elements of $\varnothing$ will be true, since no element exists to falsify our assertion.

Disjunction

Given two logical variables, M and N, their disjunction is symbolized by the connective $\vee$, as $M \vee N$, which in ordinary language reads: Today is Saturday or it is cold. In set theory, the Venn diagram for disjunction gives the union of the two sets (Fig. 158). The union of M′ and N′, $M' \cup N'$, is the set of all days that are Saturday or cold, thus making $M \vee N$ true.

Such an operation is akin to addition in arithmetic. In logic it is called non-exclusive disjunction. Sometimes ordinary language uses "or" in the exclusive sense, meaning one or the other but not both. Latin was more precise, with its use of "vel . . . vel" for "one or the other, or both", and "aut . . . aut" for "either one or the other, but not both."

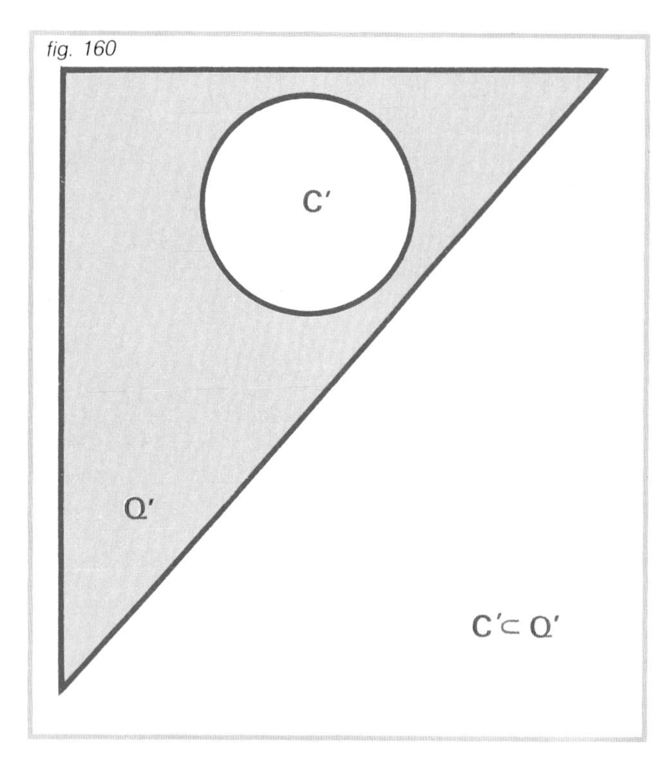

fig. 160

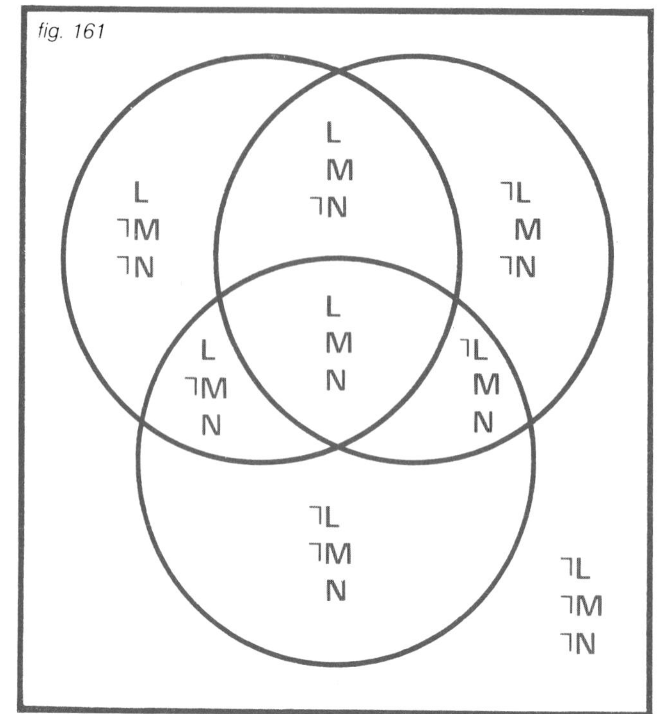

fig. 161

Graphically, the exclusive "or" is shown in Fig. 159. The symbol for this type of union is $\dot\cup$, and $M' \dot\cup N'$ is the set of all days that are Saturday or cold, except cold Saturdays. In propositional calculus, the corresponding symbol for disjunction is $\dot\vee$, and $M \dot\vee N$ is the compound proposition: Today is Saturday or it is cold, but not both.

There is yet another type of "or" in mathematical logic, namely incompatibility, which is discussed a bit later (p. 122).

Implication

As we saw earlier, implication is symbolized by an arrow between two logical variables. It corresponds to the "if . . . , then . . ." formation in everyday language and is seen graphically in Fig. 160.

Let C′ be the set of all horses and Q′ the set of all quadrupeds. The diagram immediately indicates that C′ is contained in Q′ which also contains other sets (giraffes, elephants, and so on). Using the set-theoretical symbol for inclusion, we write $C' \subset Q'$. In propositional terms this says, "All horses are quadrupeds."; in implicational terms it says, "If this animal is a horse, then it is a quadruped."

In set theory, inclusion is considered distinct from proper inclusion. C′ is properly included in Q′ if all the members of the first are members of the second, but at least one of the second does not belong to the first. Failing this, we can say that C′ is a subset of Q′, or is included in Q′, or $C' \subseteq Q'$. If two sets are identical because they have the same elements, we write $A' = B'$. Two sets containing the same number of elements are called equipotent, of the same "power."

Membership of an element in a set is symbolized by $\in$. This concept is different from being a subset which is written $\subset$. The set of horses is a subset, not a member, of the set of quadrupeds. Any horse is a member of the set of all horses, but a horse is not a subset of all quadrupeds (although the set containing only that horse is).

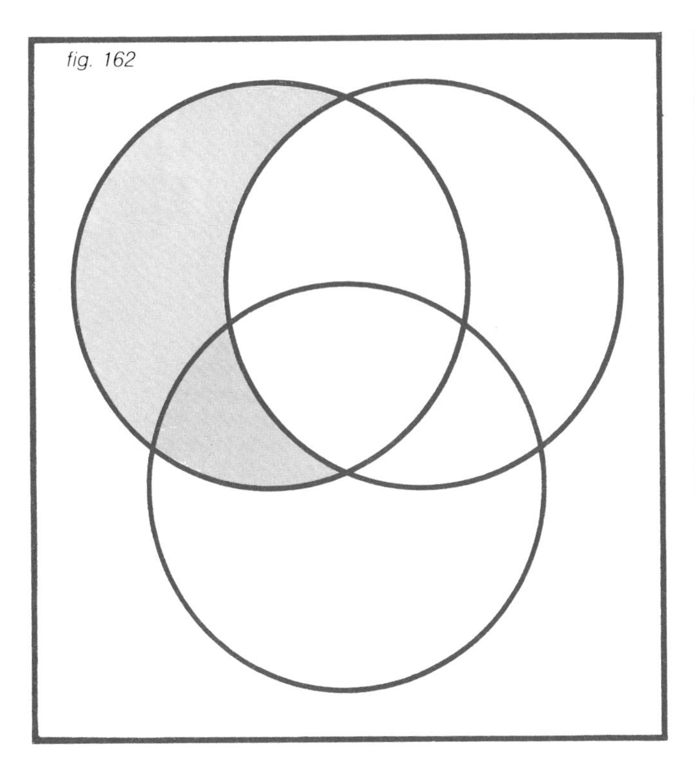
fig. 162

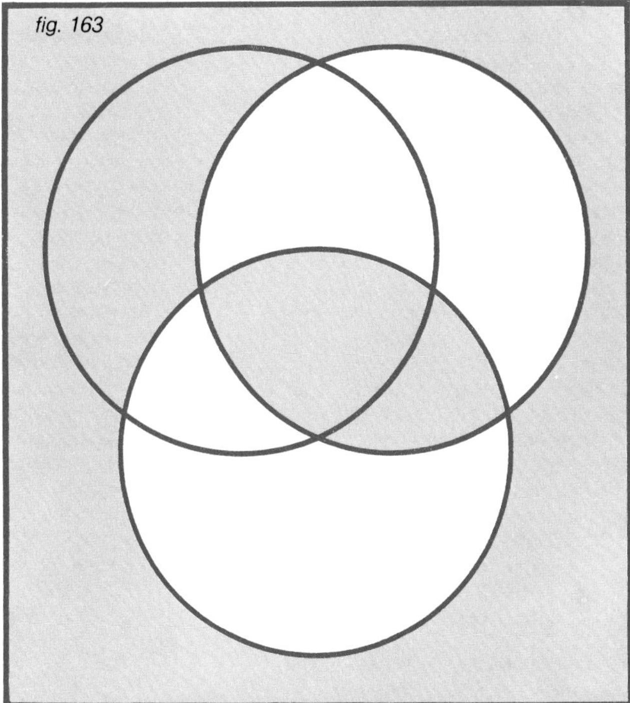
fig. 163

Who has drunk the brandy?

Venn diagrams can be used for a variety of games. Take the following riddle: Len, Mark and Nick often dine together, but we do not know who has brandy after dinner. However, we are told that:

A) If Len orders a brandy so will Mark;

B) Mark or Nick will sometimes order a brandy, but they will never order brandies at the same time;

C) Len or Nick will sometimes order a brandy, either one or both of them;

D) If Nick orders a brandy, so will Len.

Let L = Len drinks the brandy, M = Mark drinks the brandy, N = Nick drinks the brandy.
Eight possibilities emerge (Fig. 161):

1) L,M,N are all false;

2) L,M,N are all true;

3) L is true and M,N are false;

4) L,M are true and N false;

5) L,N are false and M true;

6) L,N are true and M false;

7) M,N are true and L false;

8) L,M are false and N true.

In light of the given information, we can now shade those areas in Fig. 161 in which our premises are false.

A) Colour the area where we have L, ˥M (Fig. 162);

B) colour the areas ˥M, ˥N and M,N (Fig. 163);

C) colour the area for ˥L, ˥N (Fig. 164);

D) colour the area for N, ˥L (Fig. 165).

Superimposing the four shaded figures we produce Fig. 166, which represents L, M, ˥N. Therefore, Len and Mark drank a brandy while Nick did

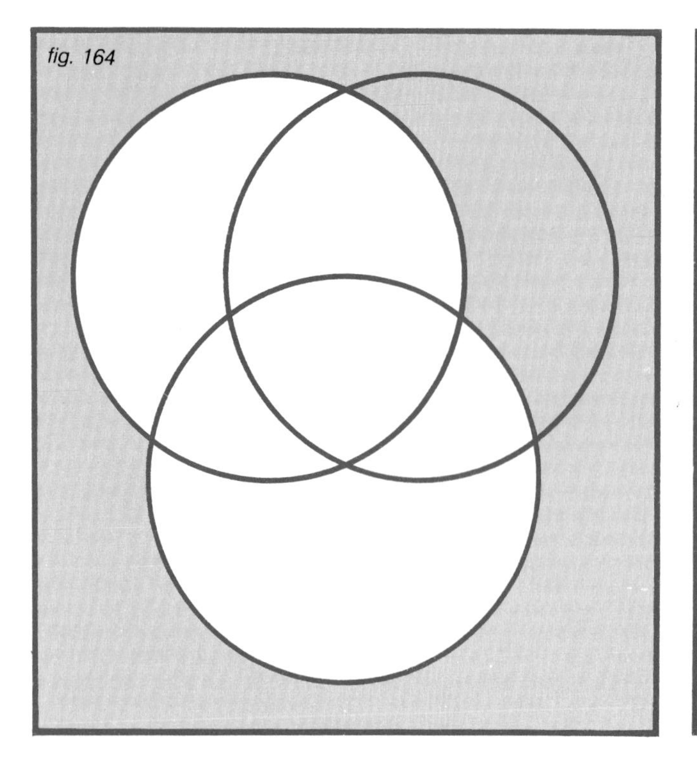
fig. 164

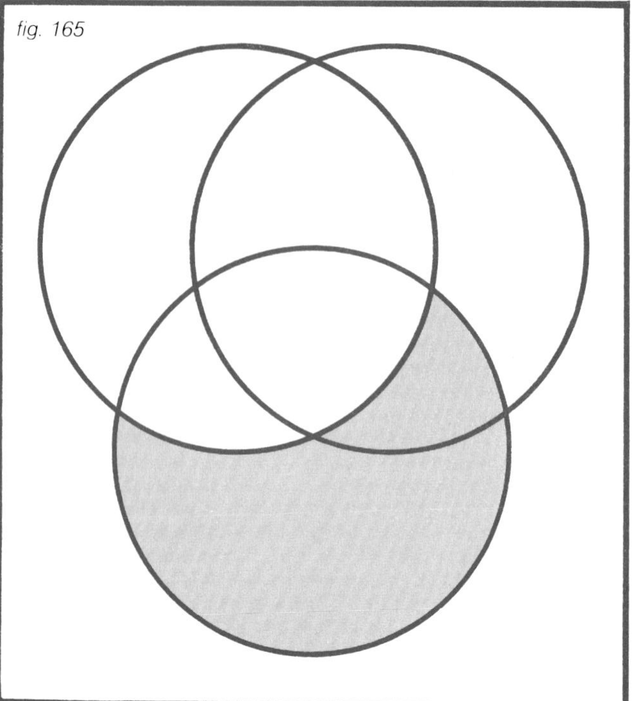
fig. 165

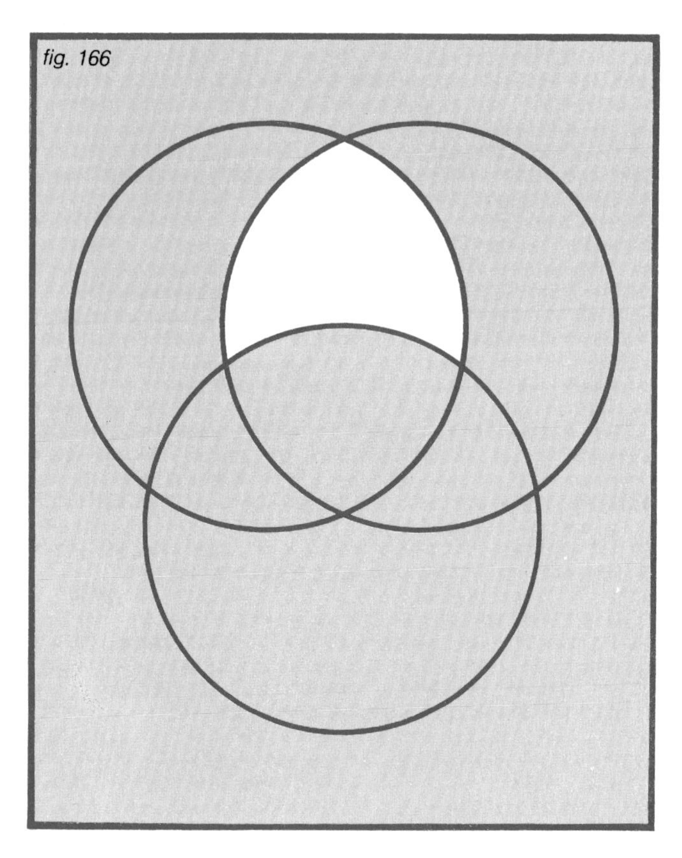
fig. 166

not. A simpler solution appears at the end of the next section.

The multiplication tables of propositional calculus: truth-tables

Turn back to Fig. 156 and the corresponding propositional expression $M \wedge N$. This proposition is true on cold Saturdays. Consider all possible truth-value combinations of M and N separately (Fig. 167), and then consider the conjunction $M \wedge N$ (Fig. 168). If M,N are both true, $M \wedge N$ is true. If M is false and N true, or M true and N false, or both false, then $M \wedge N$ is false. This is summarized in the truth-table of Fig. 168. The outer columns give all the possible combinations of 0 and 1 for the propositions M and N, while the central column gives the corresponding values for the conjunction $M \wedge N$.

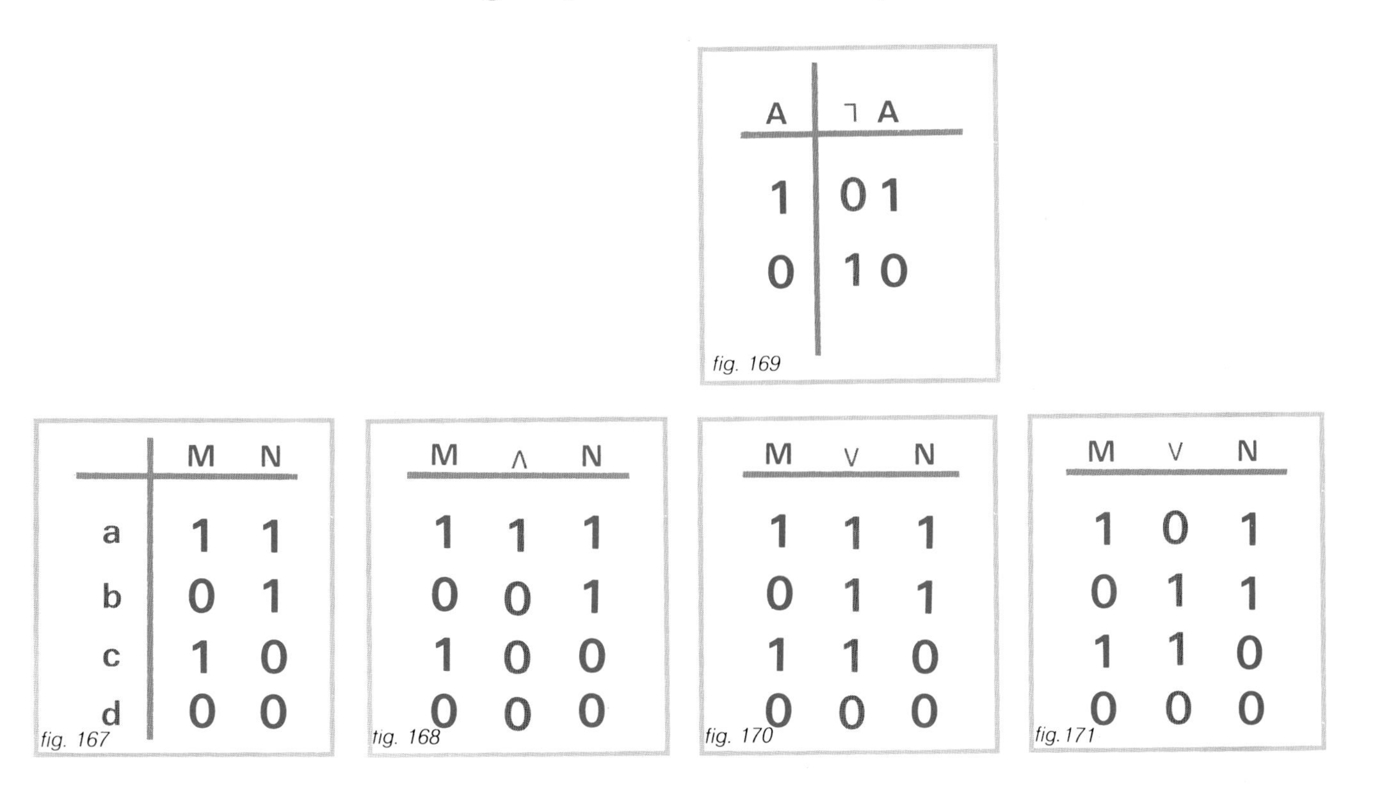

A	¬	A
1	0	1
0	1	0

fig. 169

	M	N
a	1	1
b	0	1
c	1	0
d	0	0

fig. 167

M	∧	N
1	1	1
0	0	1
1	0	0
0	0	0

fig. 168

M	∨	N
1	1	1
0	1	1
1	1	0
0	0	0

fig. 170

M	∨	N
1	0	1
0	1	1
1	1	0
0	0	0

fig. 171

Truth-tables are precise statements of the meaning of propositional connectives. Which of the following are true, and which false?

A) John has a car and the Moon is rectangular.

B) The Hudson is a river and Manhattan is an island.

C) In spring allergies occur and seagulls nest in winter.

Only B) is true; in A) and C) at least one of the conjuncts is false, hence the truth-table shows the conjunction is false. One might assert that John's having a car has nothing to do with the shape of the Moon. True, but logic is not concerned with the content of propositions, only with their truth-values. We shall deal with this later in connection with implication.

The truth-table for negation is shown in Fig. 169. For other connectives or combinations of connectives, truth-tables must be worked out step-by-step. Take the connective "or," called disjunction, which generates compound propositions. If I say, "Today is Saturday or it is cold," I could mean that each disjunct excludes the other, or I could mean that at least one disjunct is true, (and therefore possibly both). The latter is non-exclusive disjunction (Fig. 158); the former, is exclusive disjunction (Fig. 159). The corresponding truth-tables are in Figs. 170 and 171: Non-exclusive disjunction is true if at least one disjunct is true, and false only if both disjuncts are false. Exclusive disjunction, on the other hand, is true only when one disjunct alone is true; otherwise it is false.

To which type of disjunction do the following belong?

A) In September a boy or girl will be born.

B) The Moon is a satellite or a planet.

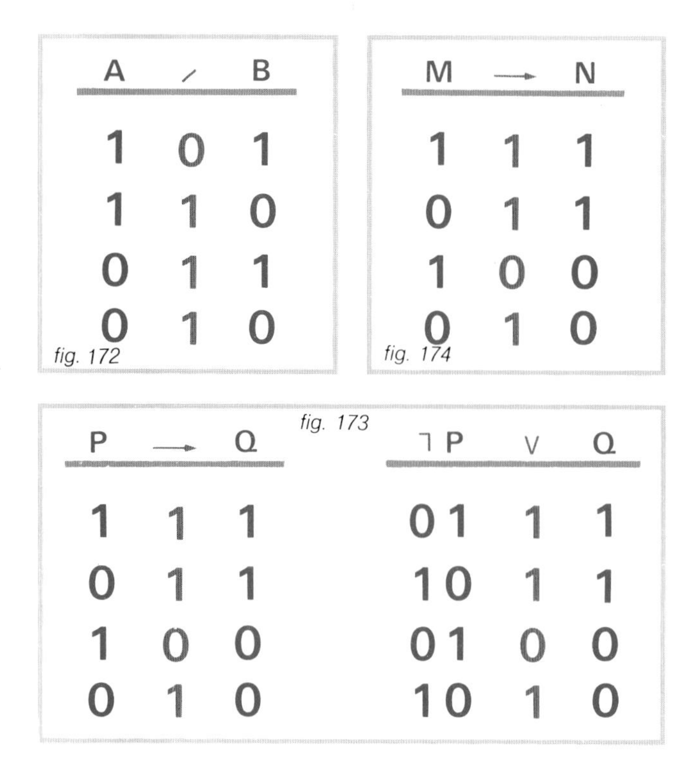

A	/	B
1	0	1
1	1	0
0	1	1
0	1	0

fig. 172

M	→	N
1	1	1
0	1	1
1	0	0
0	1	0

fig. 174

P	→	Q
1	1	1
0	1	1
1	0	0
0	1	0

¬ P	∨	Q
01	1	1
10	1	1
01	0	0
10	1	0

fig. 173

C) If you eat out or go to the movies too often, you will be broke.
The first two are clearly exclusive, while the third is non-exclusive: if you do both you will be even more broke.

Now consider the proposition, "At table a well-bred person either eats or talks." Here, a person who both ate and talked would be considered ill-bred, but one who neither ate nor talked would not necessarily be thought ill-bred. Therefore, this "or" excludes the fact that the compound is true only if both components are true. This is the "or" of incompatibility, symbolized by /. Its truth-table is shown in Fig. 172. Next, look at the truth-tables in Fig. 173. We see that the columns under the main connective are identical, which means the two compound propositions are logically identical. Moreover, it shows in what sense the arguments on p. 105 are wrong. Let us first examine the truth-table for implication, Fig. 174. Implication is false only if the antecedent is true and the consequent false; it is true in all other cases. This is not easy to grasp, especially when we consider the cases where the implication is true, but it must be remembered that logic differs from ordinary language with its many shades of meaning so dependent upon context, tone of voice, relationship between the speakers, and so on.

Take a concrete example. If S = Today is Saturday and K = I am going to the movies, then S→K means, "If today is Saturday, then I am going to the movies." The table of Fig. 175 tells us that if today is Saturday, then I am going to the movies, while if it is not Saturday, I may or may not go. We say that S is a sufficient condition for K, but not a necessary one. Of course, component propositions of an implication need not be linked at all. For example, look at, "If today is Saturday, then dogs are quadrupeds," in which the components are completely unconnected. In this instance, the meaning of the implication is clearly given by the truth-table. The compound proposition simply means that if S ("Today is Saturday.") is true, then we also get the additional true information Q ("Dogs are quadrupeds."). If both S and Q are true, so too is S→Q, while if S is true and Q false, then S→Q is false. What do we say about implication when S is false and Q true, or when both S and Q are false? Knowing that in propositional

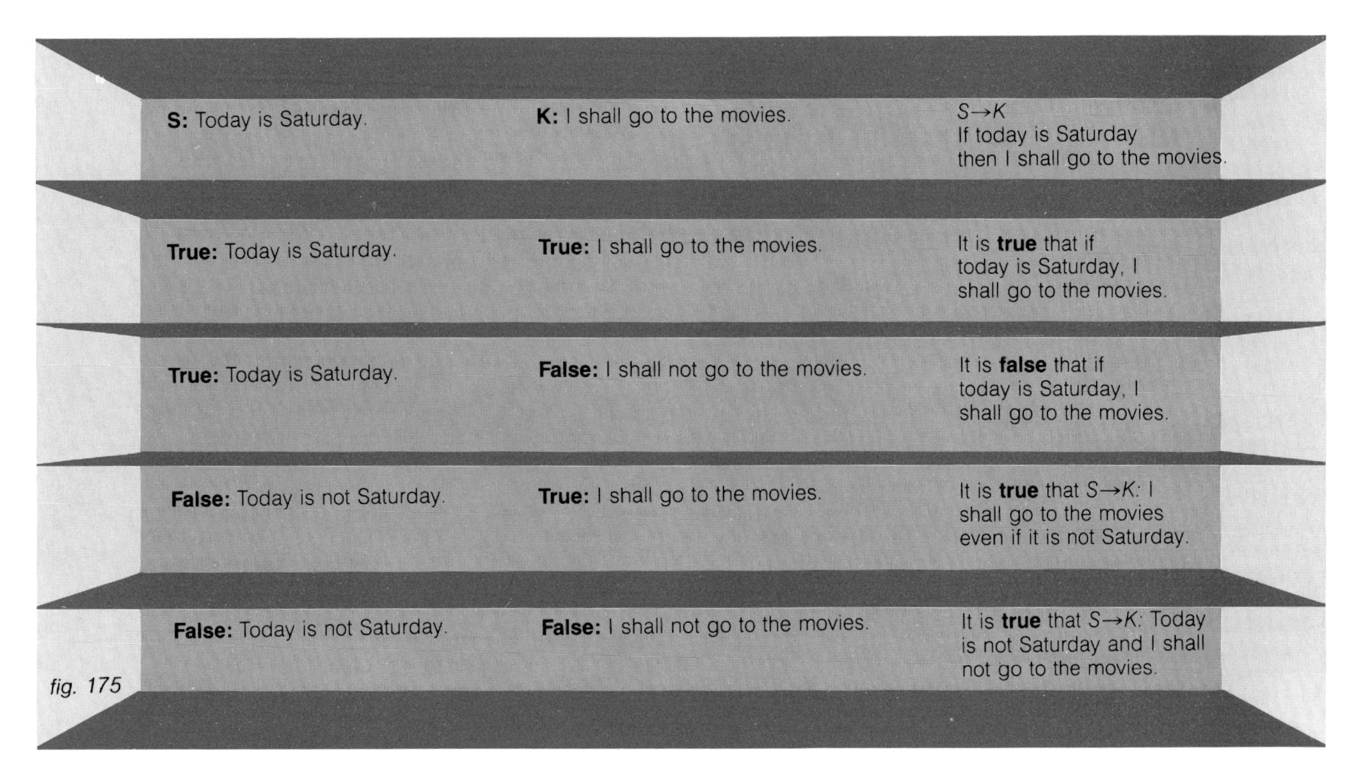

fig. 175

calculus, a proposition is either true or false but never both (law of excluded middle), we must be able to state whether an implication with a false antecedent is true or false. Now if S is false, it does not matter if Q is true or false, so in these two cases we can state $S \rightarrow Q$ to be true. In these terms, then, implication is equivalent to non-exclusive disjunction, therefore it is wise not to limit the truth-table of implication only to cases where the antecedent is true. As we mentioned earlier, medieval logicians used to say, "ex absurdis sequitur quodlibet" (from the absurd anything follows). Thus, if an antecedent is false anything can be derived: If S is false, any other proposition Q can follow, and therefore $\neg Q$ as well, which makes $S \rightarrow Q$ true. For example, both of the following are true: "If Napoleon was German, the Moon is a satellite"; "If Napoleon was German, the Moon is not a satellite." What offends common sense is that from a false antecedent we can derive a conclusion and the validity of that conclusion does not matter. In logic it is inap-

S	↔	K
1	1	1
1	0	0
0	0	1
0	1	0

fig. 176

S	↔	K	(S	⟶	K)	∧	(K	⟶	S)
1	1	1	1	1	1	1	1	1	1
1	0	0	1	0	0	0	0	1	1
0	0	1	0	1	1	0	1	0	0
0	1	0	0	1	0	1	0	1	0

fig. 177

P	⟶	Q	˥P	˥Q
1	1	1	01	01
0	1	1	10	01
1	0	0	01	10
0	1	0	10	10

fig. 178

P	⟶	Q	˥Q	˥P
1	1	1	01	01
0	1	1	01	10
1	0	0	10	01
0	1	0	10	10

fig. 178b

propriate to demand that false antecedents be excluded. Not only would it cause pointless complications but there is nothing contradictory in accepting implications with false antecedents as true. Remember, absence of contradiction is all that matters in logic.

A major difficulty in understanding the truth-table of implication is that in everyday language the connective "if . . . , then . . . " is used only when the antecedent is true. When I say, "If today is Saturday I am going to the movies," I imply that if it is not Saturday, I will not go to the movies; that is, S is also a necessary condition for K. In logic and mathematics, we must make clear and strict distinctions between the meanings of operations. When S is both necessary and sufficient for K we have equivalence, expressed by "if and only if" (often abbreviated as "iff"). The corresponding symbol is ≡, or ↔, also known as biconditional (each component implies the other). This often occurs, especially in mathematical deductions; it indicates the necessary and sufficient conditions for a given mathematical relation to exist. Its truth-table is shown in Fig. 176: equivalence holds only if both components are true or both are false.

In everyday language we often use such expressions as: "I shall go to the show if, and only if, I have a ticket." Or we might say: "If I have a ticket, then, and only then, can I go to the show"; this compound proposition is true when both components are true.

Earlier we discussed logically identical propositions. We saw that $P \rightarrow Q$ and $\neg P \vee Q$ are logically identical and can be formally expressed as $(P \rightarrow Q) \leftrightarrow (\neg P \vee Q)$ (Fig. 173). Similarly, the proposition $S \leftrightarrow K$ is logically identical to the compound proposition $(S \rightarrow K) \wedge (K \rightarrow S)$ as shown by the truth-tables of Fig. 177. Truth-tables serve to verify the validity of the conclusion of a form of argument. We enter premises and conclusion in an appropriate truth-table and make certain that whenever the premises are true, the conclusion is

$L \longrightarrow M$

$M \veebar N$

$L \vee N$

$N \longrightarrow L$

fig. 179

L	→	M	M	⊻	N	L	∨	N	N	→	L
1	1	1	1	0	1	1	1	1	1	1	1
0	1	1	0	1	1	0	1	1	0	1	1
1	0	0	1	1	0	1	1	0	1	0	0
0	1	0	0	0	0	0	0	0	0	1	0

fig. 180

also. A single line of true premises with a false conclusion indicates the form is invalid.

The argument P→Q, ⌉P,.·.⌉Q of p. 105 (illustrated earlier in Fig. 153), is examined in Fig. 178. In the second line, the two premises are true but the conclusion is false, therefore the argument is invalid. However, the argument P→Q, ⌉Q,.·.⌉P is valid, for the only time that both premises are true (line 4), the conclusion also proves true. All the other arguments can be tested in a similar fashion.

Another solution to the problem of the brandy drinkers

Truth-tables can be used to solve a variety of puzzles. Let us return to the brandy drinkers' puzzle and check the solution reached with Venn diagrams (pp. 119–120), with truth-tables. The four premises, which must be accepted as true, are shown in Fig. 179, and their corresponding truth-tables in Fig. 180. Assuming L is true, we then see that line 1 in the first truth-table indicates M is true, and that line 3 of the second truth-table indicates N is false, which is compatible with line 3 of the third truth-table and also line 2 of the fourth truth-table. Thus L, M, ⌉N is a solution. Next, assuming M is true, we see that line 3 of the second table shows N is false, and line 3 of the third table shows L is true, which is compatible with line 2 of the fourth table and line 1 of the first table. Again, L,M,⌉N is a solution. Finally, assuming N is true, we find that line 2 of the second table shows M is false and line 4 of the first table shows L is false, which is compatible with line 2 of the third table, but contradicts line 3 of the fourth table. Hence, N must be false, or ⌉N true, which again yields the solution L,M,⌉N.

Of course, it is not obvious that such a problem has a solution at all; we have demonstrated only that the present example has one. However, if the

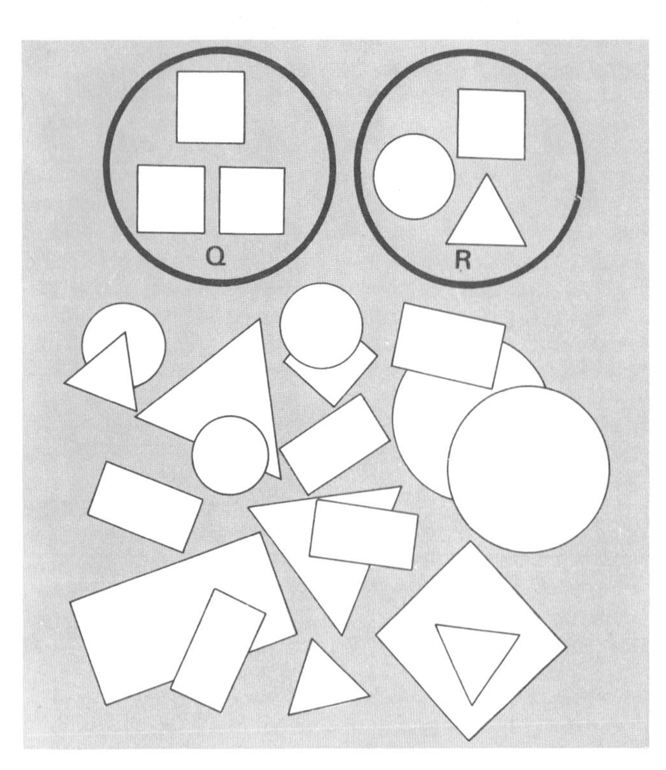

Set theory is the basis of 20th-century mathematics, and its concepts, such as those of relation and function, occur in all branches of the science. The method is indeed a sound one for teaching, even if this is not universally recognized. Although the concept is implicit everywhere in mathematical developments, it was not considered basic until the late 19th century. After numerous attempts to find a logically satisfactory and non-circular definition of the concepts "set" and "element of a set," these notions were assumed to be primitive, or undefinable. Georg Cantor's classical theory of sets (cf. p. 102) held that to define a set it was enough to describe its elements, or to refer to a criterion or a rule by which to recognize the elements. The "naive" set theory stated that each property had a corresponding set consisting of precisely those elements verifying the property. Hence the name "naive." This principle of unlimited abstraction led to antinomies and was later abandoned. The question remains unresolved, for a set may be any collection of objects, even if unrelated, which are simply put together. Two criteria for recognizing the elements are shown here: *Q* indicates the set of squares and *R* the set of red objects, namely, shape and colour respectively.

premises of a problem are themselves incompatible, there is no solution. An even easier way to solve the puzzle is to construct an eight-line table for L,M,N and the four hypotheses. The only line on which all four hypotheses emerge as true is the line for which we have L,M,$\neg$N. We invite the reader to create this table.

Who is the liar?

Truth-tables can frequently help us solve problems which seem quite difficult at first. It is important, therefore, to translate the problem into propositional form so a truth-table can be drawn. Take this example:

A boy and a girl, whose names we do not know, are sitting next to each other. One of them has fair hair, and the other dark hair. The dark-haired one says, "I am a girl." The fair-haired one says, "I am a boy." We are told one of them is lying. Which one? Let N = I am a boy, and B = I am a girl. Since one of the speakers lies, they cannot both speak the truth at the same time. Hence the premise: "It is false that N and B are true together," or, in

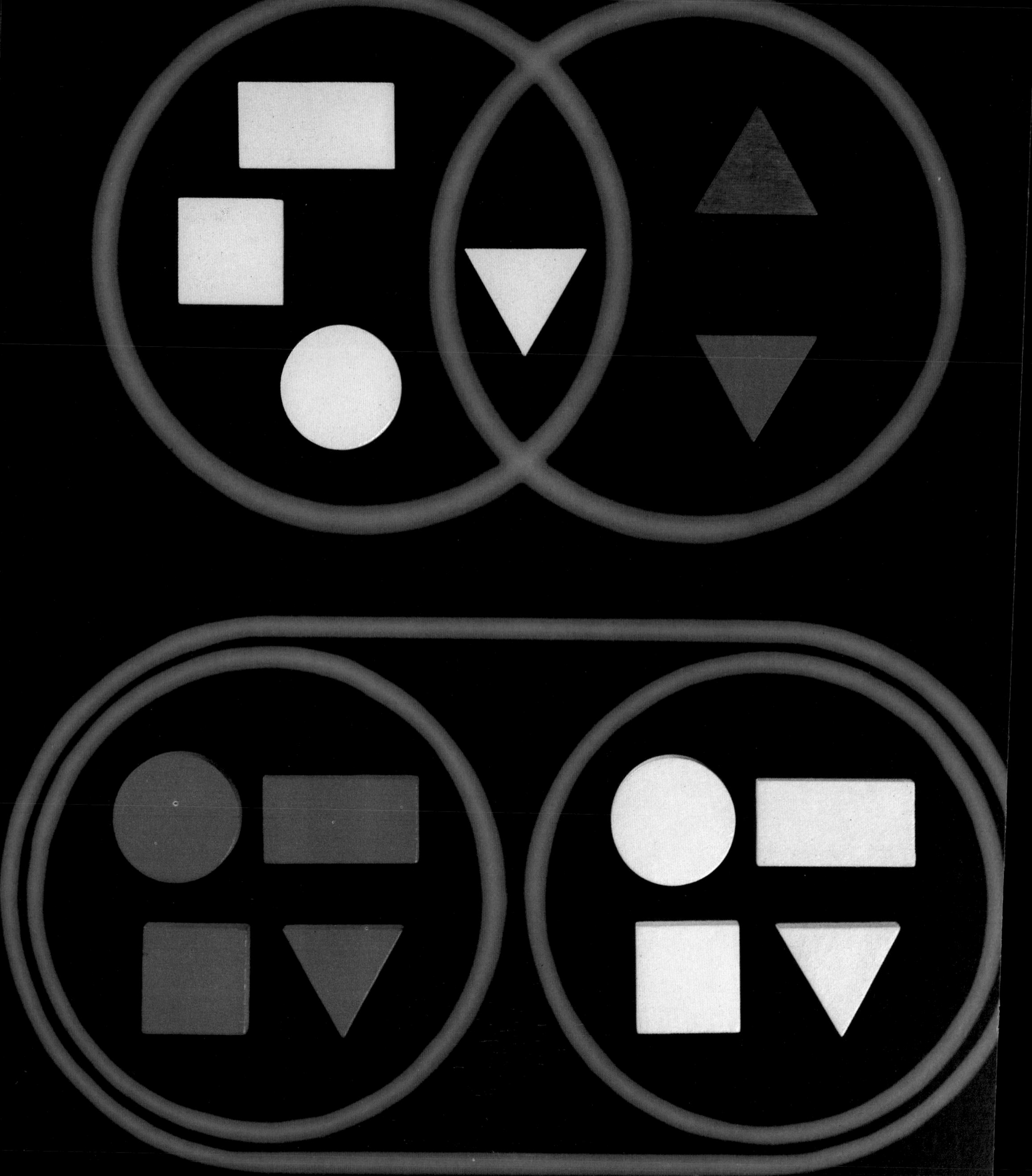

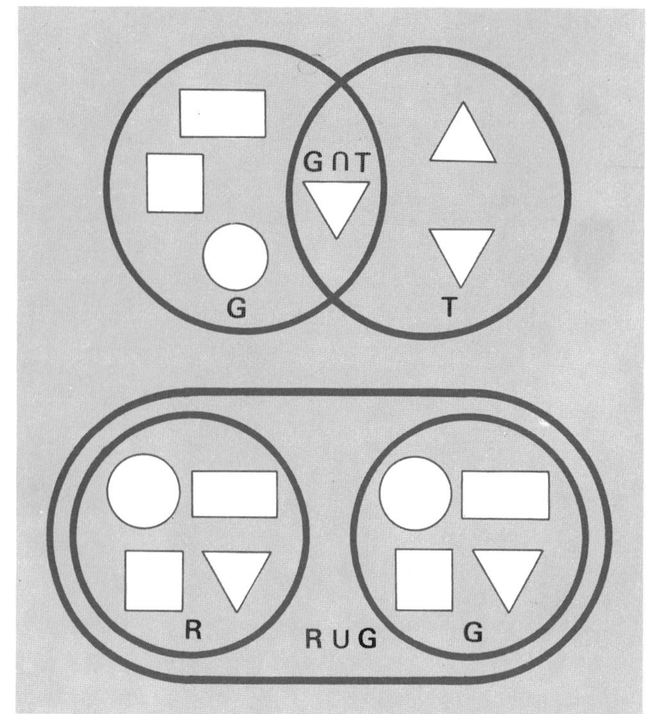

Opposite: The intersection and union of sets, shown by the block diagrams of Zoltan P. Dienes, mathematician and psychologist at Australia's University of Adelaide.
Top: The intersection of the set *G* of yellow objects and the set *T* of triangular objects, that is $G \cap T$. In this case, the only object at once yellow and triangular is the object in both circles.
Below: The union of *R*, the set of red objects, and *G*; this is $R \cup G$, whose elements are yellow or red.

symbols, $\neg(N \wedge B)$. The corresponding truth-table is shown in Fig. 181. We see from it that case a) must be excluded. Cases b) and c), where only one of them is lying and the other one is telling the truth, cannot actually occur because if we accept that one is lying, then the other, who is stating the opposite, cannot be telling the truth either. The true case can only be the one where both lie, namely d). Therefore the dark-haired one must be the boy and the fair-haired one the girl.

	¬	(N	∧	B)
a)	0	1	1	1
b)	1	1	0	0
c)	1	0	0	1
d)	1	0	0	0

fig. 181

How to argue by diagram

So far we have used truth-tables to check whether arguments are valid. The same can be done with simple diagrams which present the situations visually. Consider the following: All cats are felines. Let G′ = the set of cats, F′ = the set of felines; the

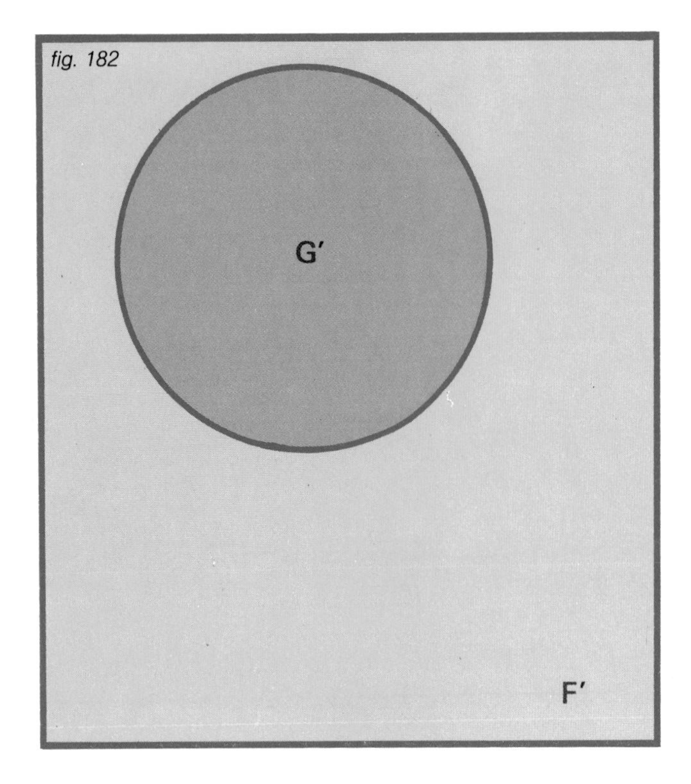

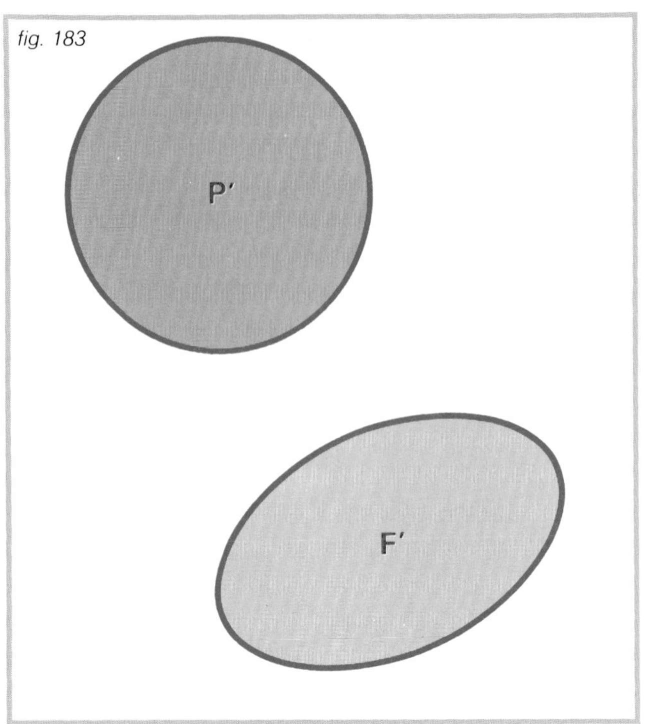

Venn diagrams then are a valid set-theoretical interpretation of the relation between cats and felines as stated in the above proposition. Fig. 182 shows this graphically. The same relation could also have been expressed by, "Some felines are cats," which states that the set of cats is included in the set of felines.

In logic, propositions such as "all A are B" are called universal affirmatives, and Fig. 182 shows how the universes of cats and felines are related. A universal proposition can be negative, as is: "No penguins are felines." Let P′ = the set of penguins and F′ = the set of felines; Fig. 183 then shows that these two sets exclude one another. Notice that this proposition is also true: "No felines are penguins," (universal negative propositions are "symmetrical"). Next take the proposition: "Some politicians are intelligent," with R′ = the set of politicians and I′ = the set of intelligent people. This produces the three cases of Figs. 184, 185, 186. Only Fig. 184 represents the proposition under consideration. Fig. 185 represents: "All intelligent people are politicians," while Fig. 186 represents: "All politicians are intelligent."

In Fig. 184, an assertion is made about only some elements of a set (only some politicians are intelligent); its corresponding proposition also concerns only some politicians. Such propositions are known as particular affirmative. The

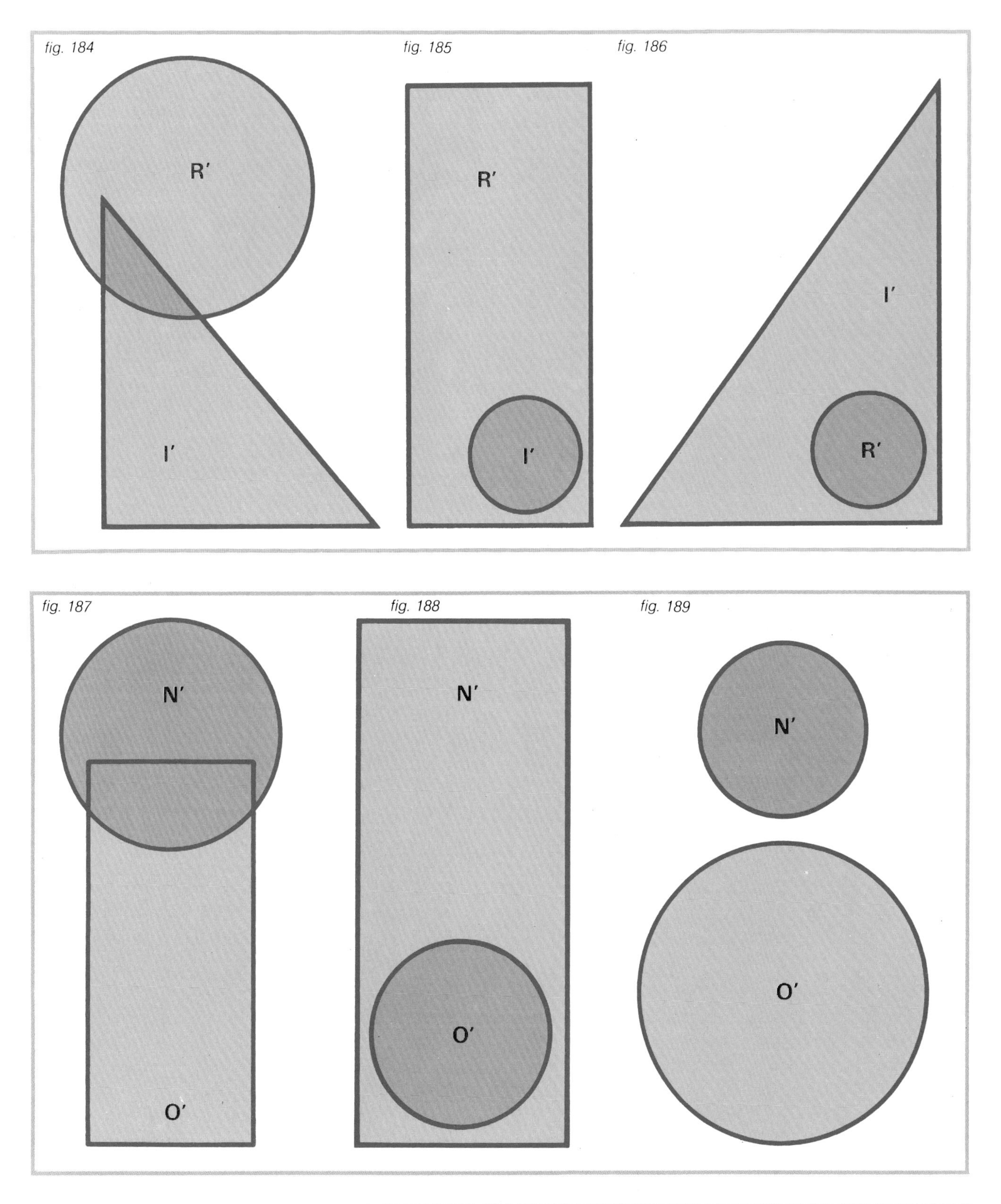
fig. 184
R′
I′
fig. 185
R′
I′
fig. 186
I′
R′
fig. 187
N′
O′
fig. 188
N′
O′
fig. 189
N′
O′

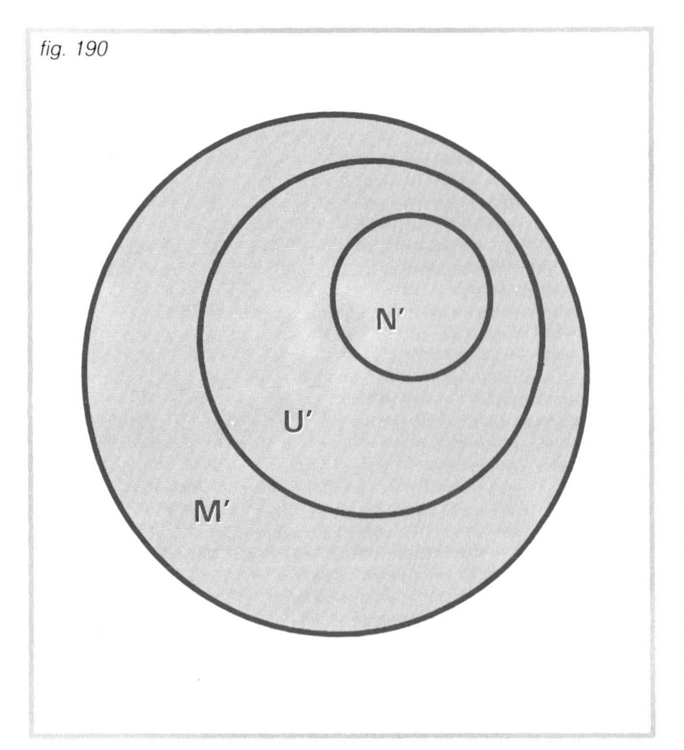

fig. 190

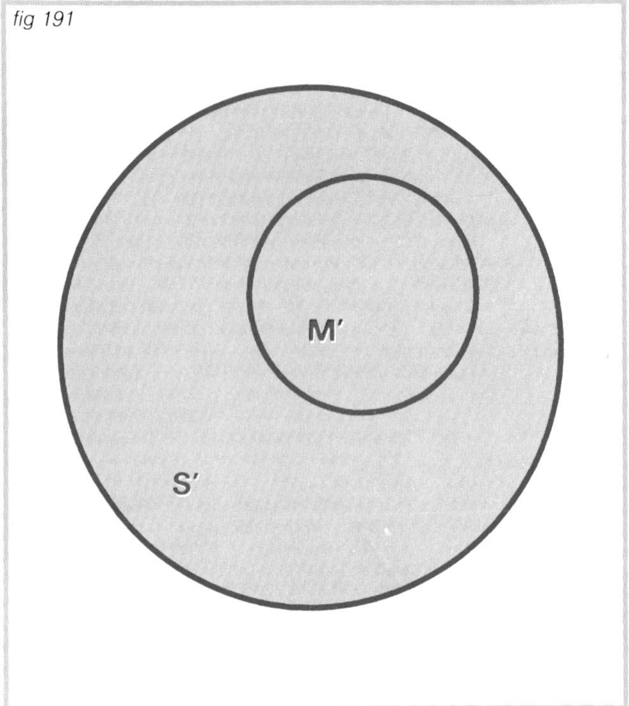

fig 191

proposition: "Some politicians are not intelligent," on the other hand, is called a particular negative. Consider a mathematical example: "Some even numbers are not divisible by 5." Let N′ = the set of even numbers and O′ = the set of even numbers divisible by 5, or more formally, $N' = \{2n\}$, $O' = \{10n\}$ where n is any integer. In Fig. 187 we see the Venn diagram. In Fig. 188 we find the true statement: "All even numbers divisible by 5 are even," while Fig. 189 is wrong because it states: "Even numbers divisible by 5 are not even."

The four forms of propositions interpreted here in terms of sets are the Aristotelian forms, named for Aristotle who first listed them. According to Aristotle, these propositions are used in scientific discourse to create the complex forms of reasoning he called syllogisms. For example: "All men are mortal, all Greeks are men, therefore all Greeks are mortal." This can be shown, as it is in Fig. 190, with M′ = the set of mortals, U′ = the set of men, N′ = the set of Greeks. Indeed, this is how we have interpreted implication. A syllogism, then, can be seen as an implication, provided we can recover the appropriate form from the expressions just considered.

A syllogism has three parts (depicted in a diagram as sets): the major premise ("All men are mortal"), the minor premise ("All Greeks are men"), and the conclusion ("All Greeks are mortal"). Each set is involved twice.

This is the simplest form of syllogism. There are other, more complicated forms whose validity is harder to determine. Again, we use Venn diagrams. Consider the following: "All students of King's College are in higher education, some people in higher education are terrorists, therefore some students of King's College are terrorists." Let M′ = the set of students at King's College, S′ = the set of students in higher education, T′ = the set of terrorists. The first premise is represented by Fig. 191, while the second, being a particular affirmative, is compatible with

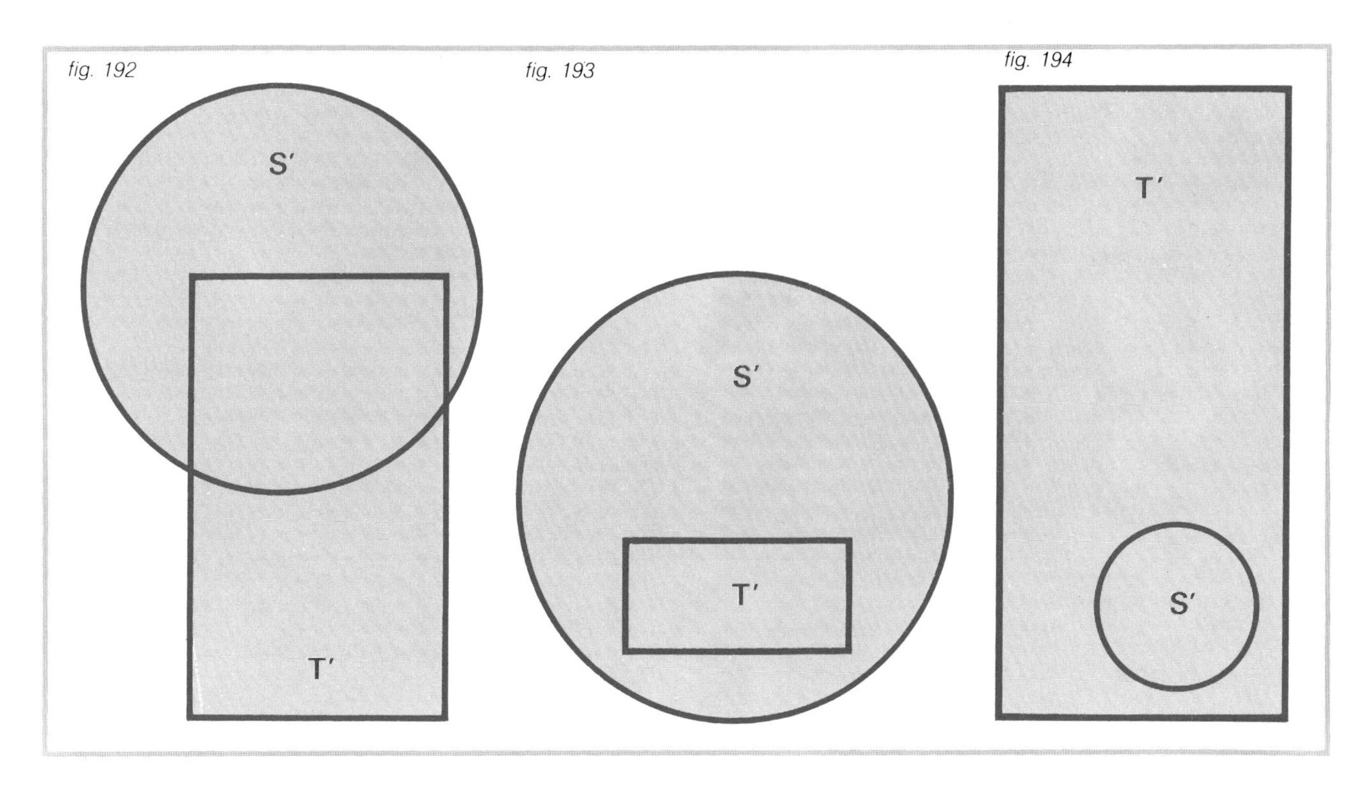

Figs. 192, 193, 194. The sets S′,M′,T′ can be arranged in eight ways (Figs. 195–202). These diagrams show that an element of T′ (a terrorist) is not always a member of M′ (a student of King's College). Indeed, this is obvious from the very first figure. Hence, the syllogism is invalid. The conclusion does not follow from the premises.

Now consider this argument: "All children have two legs, all chickens have two legs, therefore all chickens are children." Let B′ = the set of children, G′ = the set of bipeds, P′ = the set of chickens. Fig. 203 represents the first premise. However, we know that the only possible representation of how the sets of children and chickens are related is given by Fig. 204; the two exclude each other (they have no common member), therefore the three sets are related as shown in Fig. 205. No member of G′ is both child and chicken. Again the syllogism's conclusion is invalid.

Let us now represent the following in diagrams:

A) "All rational numbers are real numbers."

B) "All integers are rational."

C) "All irrationals are real."

D) "No integer is irrational."

The results are shown in Fig. 206 (cf. p. 136).

Finally, consider: "No man is an aquatic animal, some aquatic animals are mammals, therefore some mammals are not human." Let U′ = the set of men, A′ = the set of aquatic animals, M′ = the set of mammals. The first premise is represented by Fig. 207 and the second is compatible with Figs. 208–210. We know that not all aquatic animals are mammals, nor are all mammals aquatic (Figs. 209, 210), nevertheless, we present these possibilities to see if our conclusion, that some mammals are not human, can be contradicted. The possible arrangements of the three sets are shown in Figs. 211–217 We see there is always

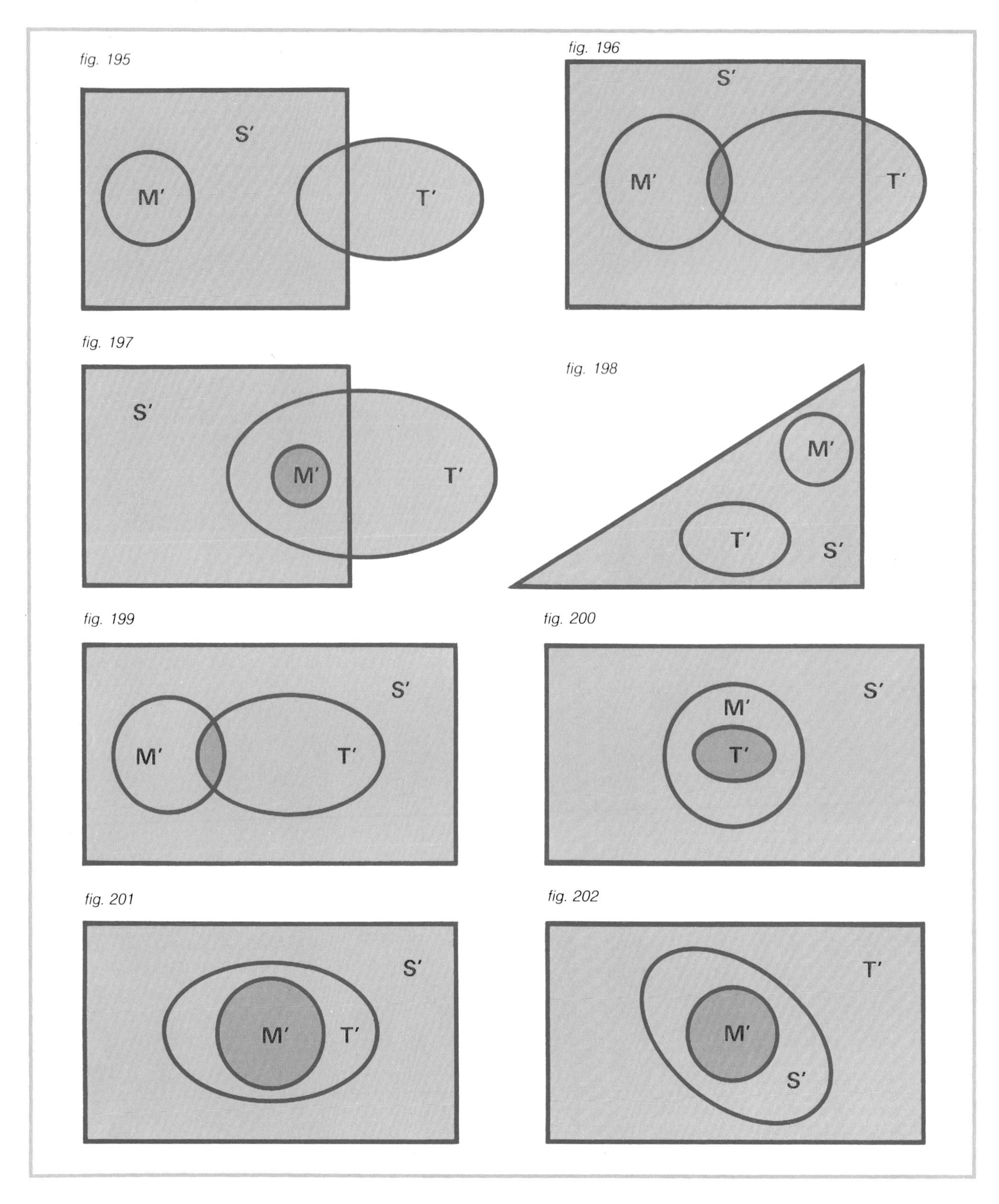
fig. 195
S′
M′
T′
fig. 196
S′
M′
T′
fig. 197
S′
M′
T′
fig. 198
M′
T′
S′
fig. 199
S′
M′
T′
fig. 200
S′
M′
T′
fig. 201
S′
M′
T′
fig. 202
T′
M′
S′

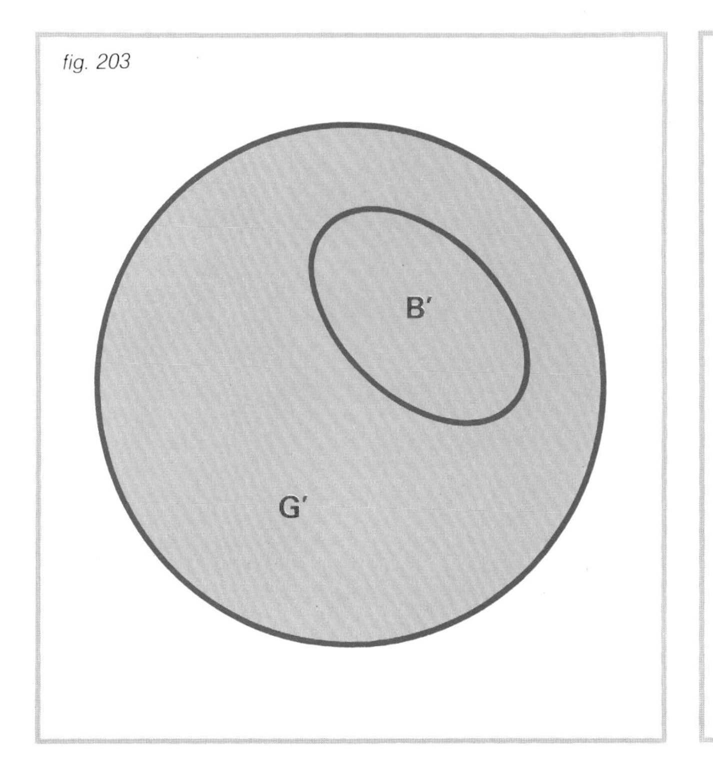

fig. 203

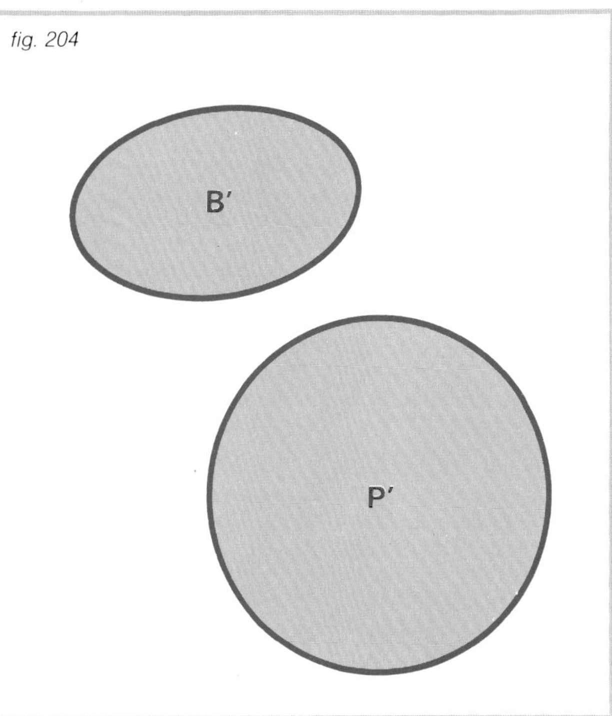

fig. 204

some element of M′ not belonging to U′, so the syllogism is valid.

A practical application: logic circuits

Recently, the propositional calculus has been used to develop modern electronic processors. Here, we shall give an intuitive account of the general principles involved in this genuine technological revolution.

Consider the electric circuit of Fig. 218, consisting of a battery p, a switch i, and a lamp l. If the switch is closed, current flows and the lamp is lit; if the switch is open, no current flows and the

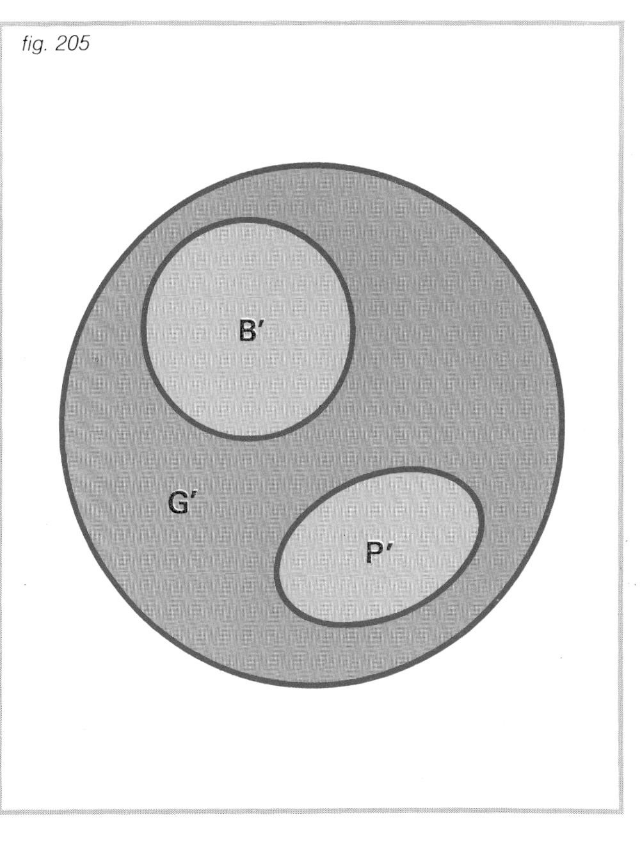

fig. 205

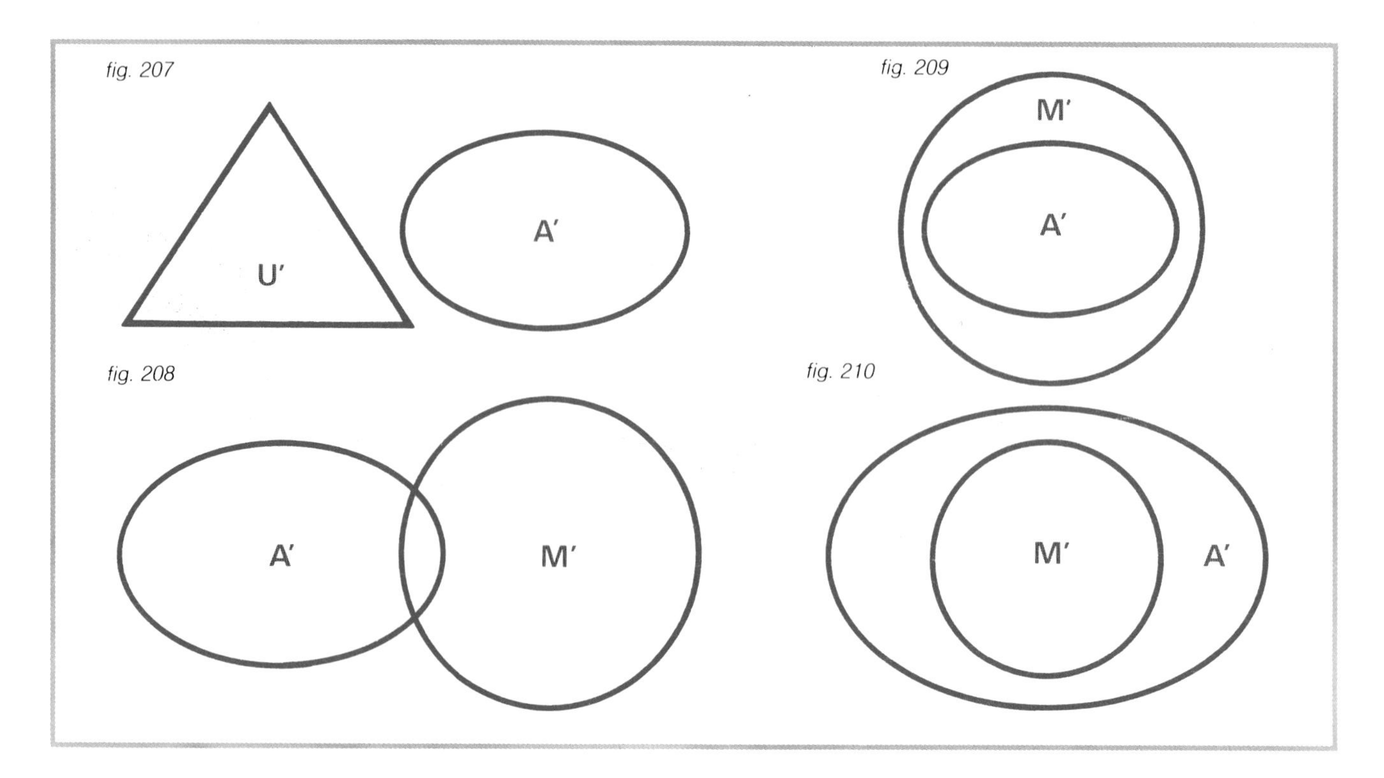

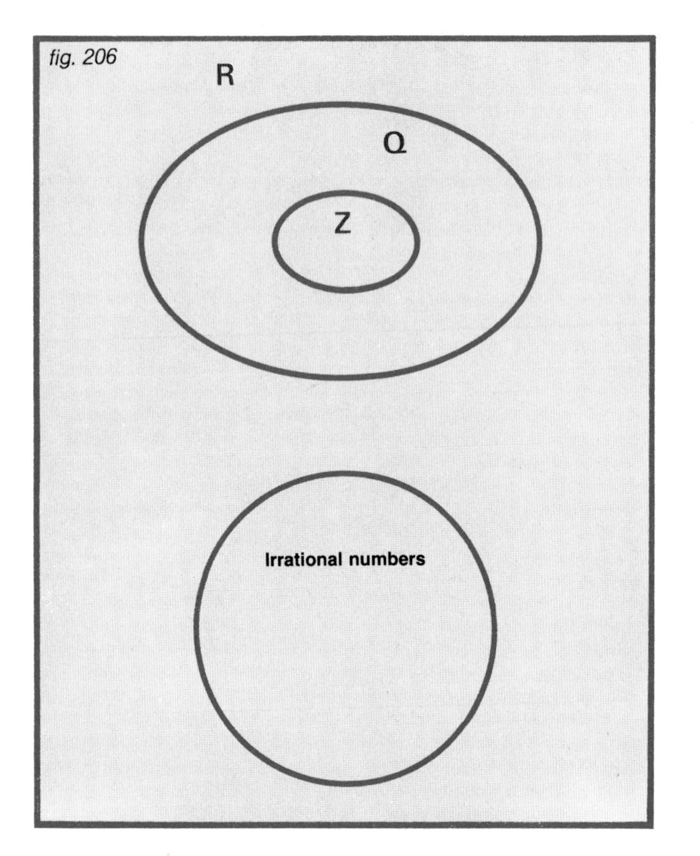

lamp remains unlit. Thus, in the circuit, current can either flow or not. The circuit, then, can be in two states, like a variable in propositional calculus (that is, 1 or 0). This can be illustrated as it is in Fig. 219, where A is a kind of proposition ("The switch is closed"), true when the switch is closed and false when the switch is open. A circuit that joins two points through the switch is a commutation circuit. With these simple examples we can now translate Boolean algebra and propositional calculus into electric circuits that perform the same logical tasks.

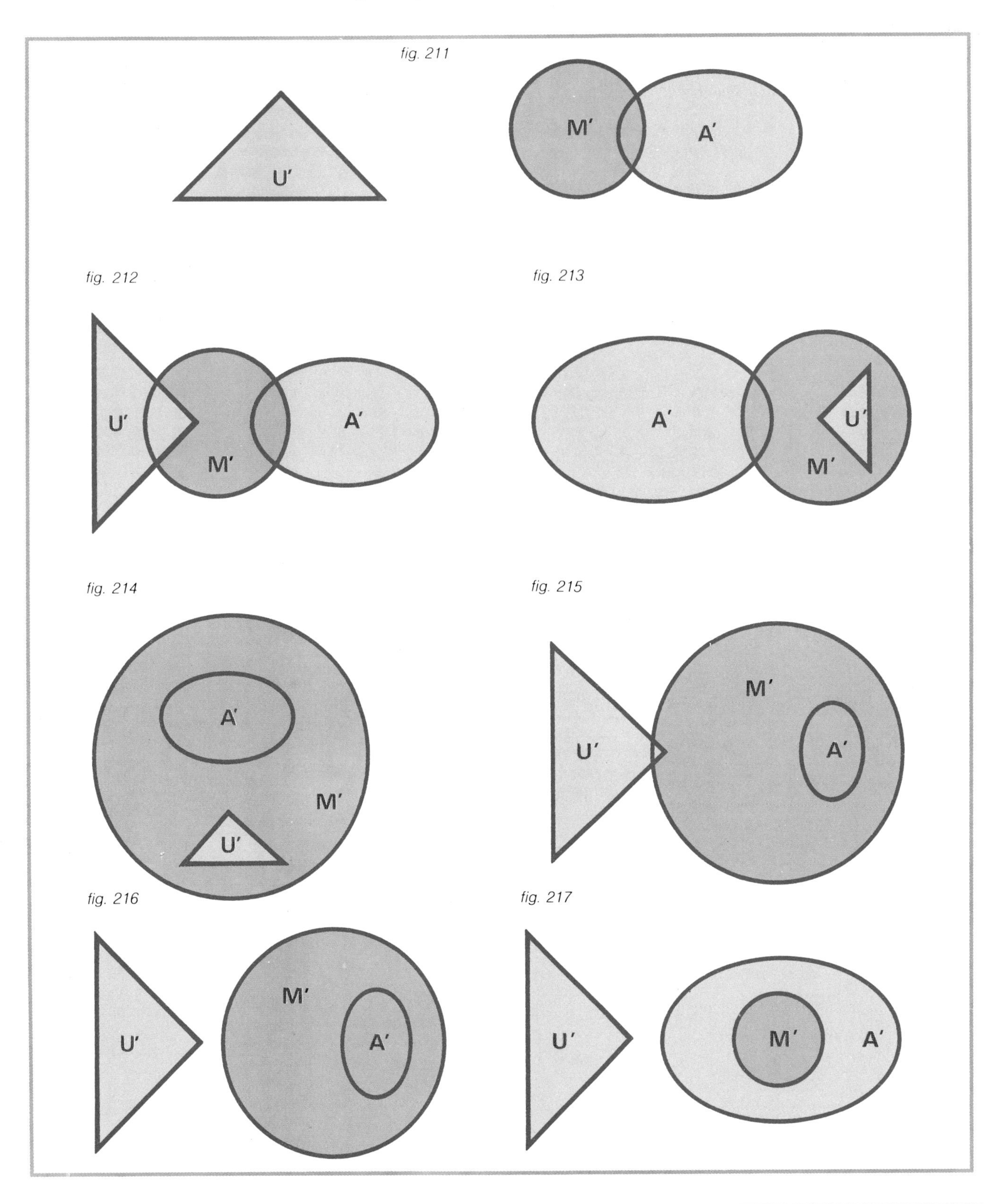

fig. 211

fig. 212

fig. 213

fig. 214

fig. 215

fig. 216

fig. 217

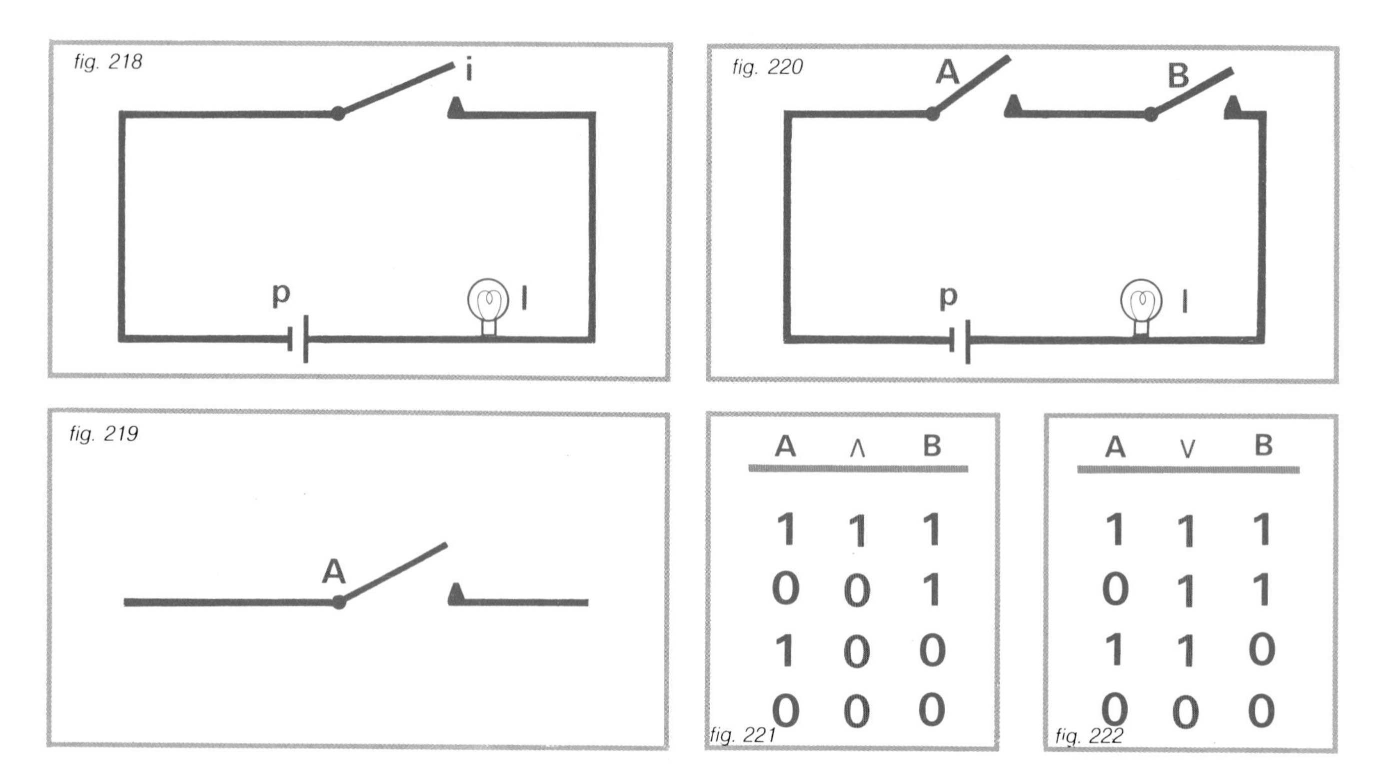

fig. 218

fig. 220

fig. 219

A	∧	B
1	1	1
0	0	1
1	0	0
0	0	0

fig. 221

A	∨	B
1	1	1
0	1	1
1	1	0
0	0	0

fig. 222

The principle is this: If for a valid argument we need only apply certain rules to premises, and if all relations between premises are summarized in a few tables, then an electronic calculator can make these deductions, that is it can calculate. Take the previous circuit and insert two switches A and B, as in Fig. 220. The lamp will be lit only when both A and B are closed. If the lamp when lit corresponds to the value 1, or true, and when unlit corresponds to the value 0, or false, then the circuit represents the operation of conjunction of A and B, Fig. 221.

What is the corresponding circuit for non-exclusive disjunction? Using the truth-table in Fig. 222, we discover that only one switch need be closed for the current to pass and the lamp to be lit. A circuit satisfying this property is seen in Fig. 223.

Next, let us try to find the commutation circuit for implication, with the truth-table in Fig. 224, which, incidentally, reminds us that $(A \rightarrow B) \leftrightarrow (\neg A \vee B)$. Hence, we only need to represent disjunction (Fig. 225). Now $\neg A$ is true exactly when A is false. If A represents "the switch is open," then

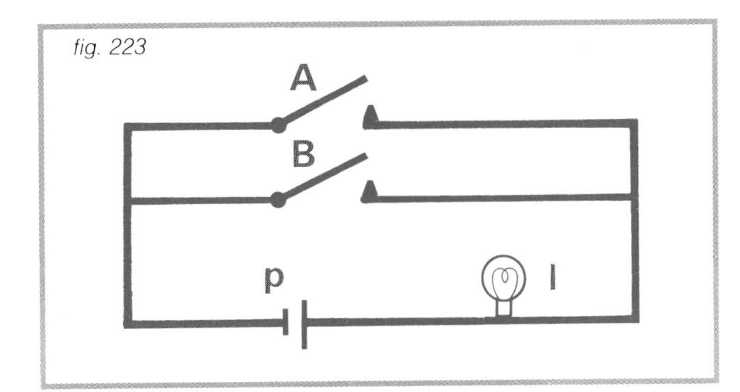

fig. 223

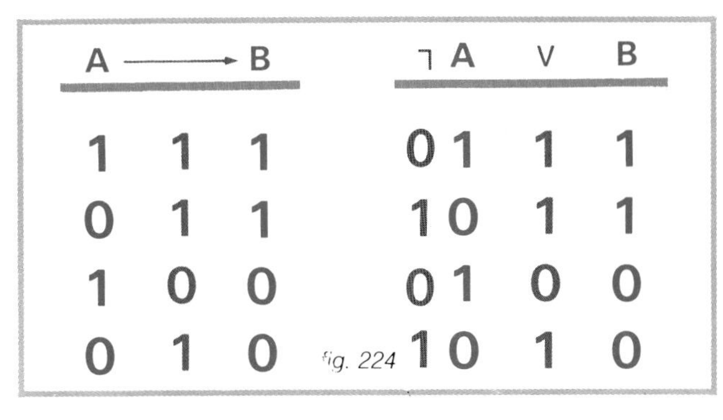

A	→	B
1	1	1
0	1	1
1	0	0
0	1	0

¬	A	∨	B
0	1	1	1
1	0	1	1
0	1	0	0
1	0	1	0

fig. 224

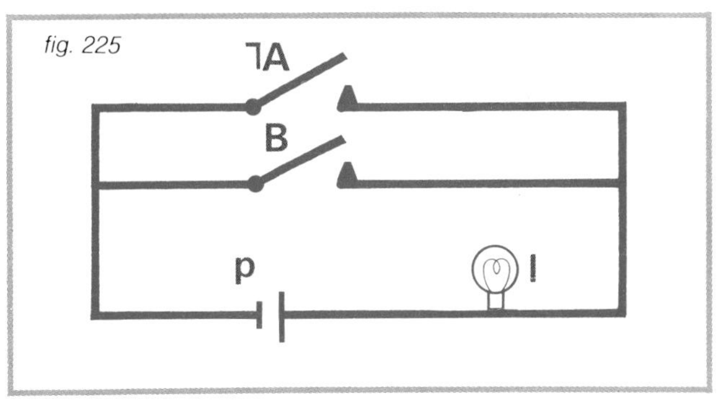

fig. 225

the lamp will be lit when ¬A is true, that is, when the switch is closed.

Here we have replaced implication by disjunction, which is simpler for us. It can be shown that conjunction, disjunction and negation suffice to represent all other logical operations.

Let us look at an interesting puzzle conceived by a Hungarian mathematician. We must construct an electric circuit for a three-bed sleeper compartment in such a way that the light stays on only when at least two of the occupants (a majority) want it on, and that an occupant who wants

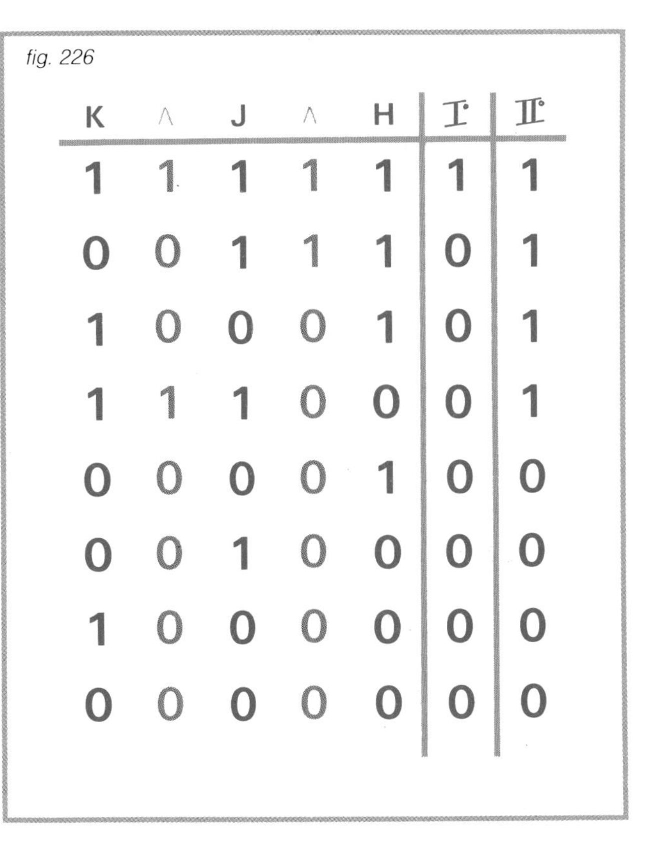

K	∧	J	∧	H	I°	II°
1	1	1	1	1	1	1
0	0	1	1	1	0	1
1	0	0	0	1	0	1
1	1	1	0	0	0	1
0	0	0	0	1	0	0
0	0	1	0	0	0	0
1	0	0	0	0	0	0
0	0	0	0	0	0	0

fig. 226

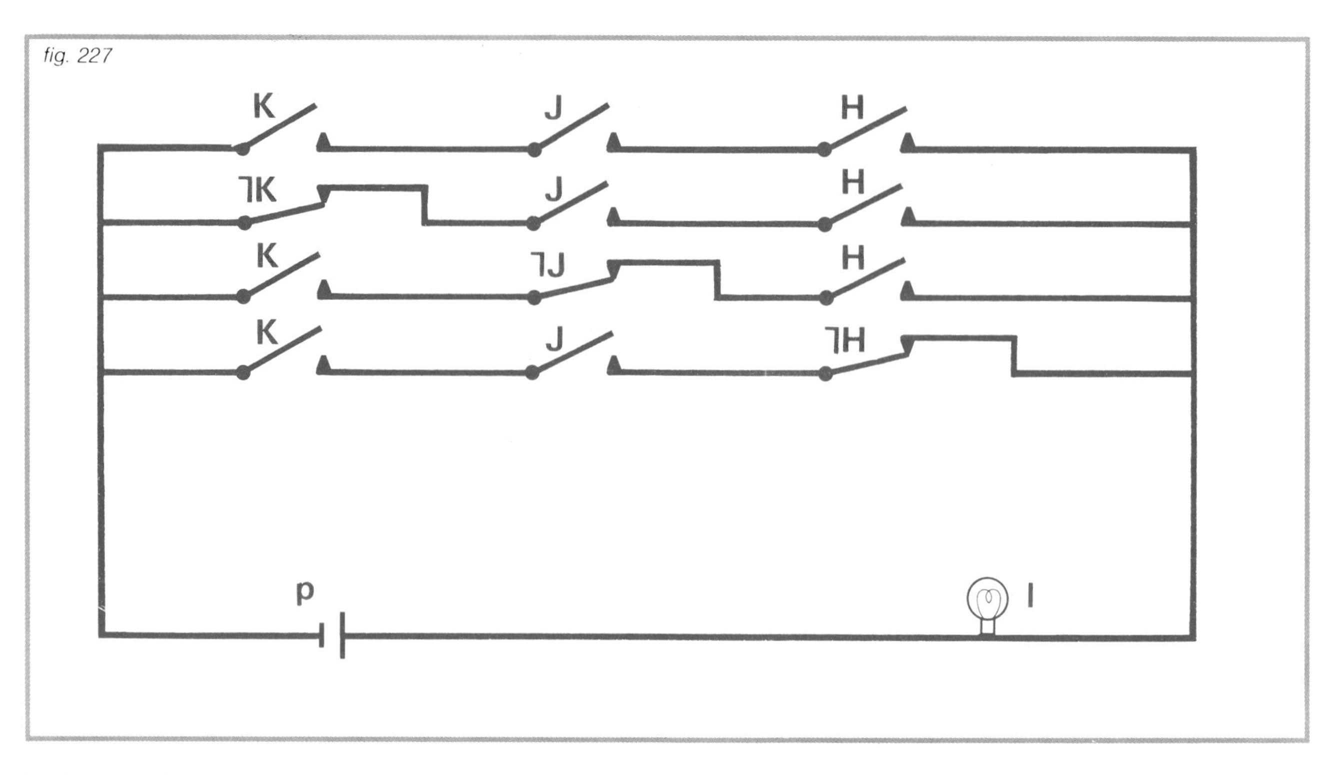

the light on has only to push a button. The circuit must therefore be planned so the light remains on in case at least two occupants have pushed their buttons. Let the occupants be K,J,H, and the letters signify their respective light buttons. Now we can construct a truth-table for the conjunction of K, J and H and consider those cases in which at least two of the variables are true.

$$(K \wedge J \wedge H) \vee (\neg K \wedge J \wedge H) \vee (K \wedge \neg J \wedge H) \vee (K \wedge J \wedge \neg H)$$

Thus, we have the formula shown here in symbols, and it reads: The lamp will be lit if K,J,H press their buttons, or if J and H do so but K does not, or if K and H do so but J does not, or if K and J do so but H does not. The circuit is shown in Fig. 227. The switches with the same letter must be connected to one another in such fashion that if an occupant presses his button all the switches with that letter will close, while all switches bearing a negation sign before that letter will open. Conversely, if an occupant does not press the button the switches with that letter remain open, while those with the negation sign before that letter remain closed.

Now let us examine a problem of an opposite type. Given the circuit of Fig. 228, what is its formula in terms of propositional calculus? We observe that the lamp will be lit if at least one of the pairs S and T, or U and V, is closed. Thus we have $(S \wedge T) \vee (U \wedge V)$.

fig. 228

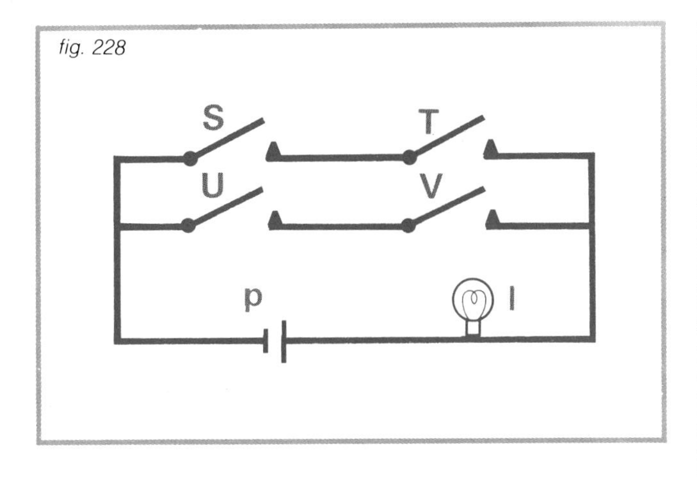

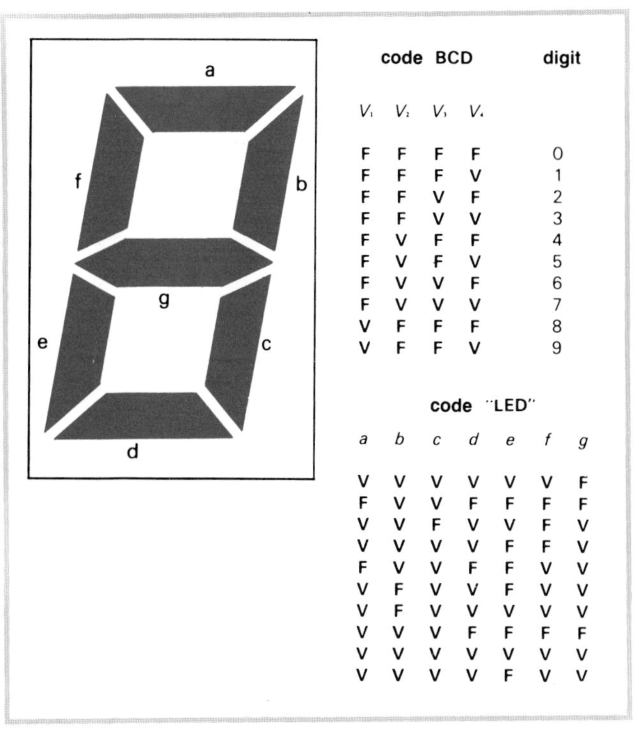

code BCD				digit
V_1	V_2	V_3	V_4	
F	F	F	F	0
F	F	F	V	1
F	F	V	F	2
F	F	V	V	3
F	V	F	F	4
F	V	F	V	5
F	V	V	F	6
F	V	V	V	7
V	F	F	F	8
V	F	F	V	9

code "LED"						
a	*b*	*c*	*d*	*e*	*f*	*g*
V	V	V	V	V	V	F
F	V	V	F	F	F	F
V	V	F	V	V	F	V
V	V	V	V	F	F	V
F	V	V	F	F	V	V
V	F	V	V	F	V	V
V	F	V	V	V	V	V
V	V	V	F	F	F	F
V	V	V	V	V	V	V
V	V	V	V	F	V	V

Electronic numerals, widely used in digital watches and pocket calculators, are evidence that some operations once thought peculiar to the human mind can be automated and performed by machines. The diagram shows an LED generator of decimal digits. Each of the seven bars, marked a to g, is a moulded light-emitting diode (LED). The codes shown in the truth-tables are used in a small electronic processor to transform electric impulses into numbers which can then be read. For example, instructed to show the number 7, a, b, c, will light up, while to produce the number 1 only b, c will light up.

GAMES WITH PROBABILITY

The most important questions in life are, for the most part, merely problems in the calculus of probability. (Pierre Simon Laplace)

The reality of chance and uncertainty

Mathematical logic applies to those propositions that are either true or false, and to situations and objects about which we can determine whether they have a certain property, or lack it. Falsehood or lack is represented by 0, and truth or presence by 1. In dealing with paradoxes and antinomies we saw that the set of unitary fractions (with numerator 1) between zero and 1 is in a one-one correspondence with the set of natural numbers (Fig. 152). 0 and 1 are the limiting terms of the set. In much the same fashion, mathematical logic—the methods and procedures of which have proven highly useful—deals with those linguistic cases of the absolutely true (1) and the absolutely false (0). With mathematical logic, the truth or falsehood of compound propositions created from simple ones by logical connectives, can be deduced automatically, making it possible to instruct a machine to work faster, and often more accurately, than the human mind. However, there are a myriad situations between true and false where mathematical logic does not apply. We do not live in a world of unconditional choices, but rather one which, more often than not, demands that our choices be made with caution and deliberation. Although we are surrounded by risk and chance, our psychological makeup is such that we tend to flee uncertainty, and are more comfortable dealing with reality in terms of absolutes.

Yet, is it not more prudent—and more realistic—to avoid absolute certainty, to accept change, to assess everything for and against, to analyze our approach, and to present our beliefs as probabilities rather than as truths or falsehoods? After all, are not risk and uncertainty more a part of the human condition than certainty?

Cards, dice, games of chance and bets: historical origins of the calculus of probability

The mathematical theory of probability began with the attempts of late 17th-century thinkers to work out the possibilities of winning or losing at games of chance, cards, dice and spinning coins.

Galileo Galilei (1564–1642), and Girolamo Cardano (1501–1576) were the first to treat certain problems of dice in terms of probability. Later Blaise Pascal (1623–1662), the French philosopher and mathematician, offered a systematic account of several problems in games of chance. Pascal and Pierre Fermat (1601–1665), are considered the founders of the mathematical probability theory. Although their correspondence in 1654 indicates that between them they established the theory's basic principles, they arrived at their conclusions independently; their proposed solutions for certain problems differ only in marginal details.

In recent years, probability theory has become an essential tool in our understanding of reality, and it is used in almost every branch of science as well as in daily life. It has been argued that all knowledge is in some measure undetermined and uncertain and therefore merely probable, and that both objective and subjective reality can be discussed only in terms of probability. The following assertions suggest the wide range of situations in which probability plays a part:

A) This child has been too exposed to the cold and will probably catch the flu.

B) The sky is overcast; it will probably rain.

C) John is a compulsive smoker and runs the risk of developing lung cancer.

D) With unemployment and social and economic inequities increasing, the crime rate is likely to rise.

E) This car is parked where it shouldn't be and the owner will probably be fined.

Before presenting some probability-based games and puzzles, we should consider a few basic concepts.

Chance phenomena

If we throw a coin in the air we know it will fall, but we do not know if heads or tails will be up. If we throw a true die it too will certainly fall, but what the face will show is uncertain. Such throws are typical chance events. In general terms, a chance phenomenon is an event that can be presented as a set of alternatives. The alternatives for the die, for example, are the faces numbered 1 to 6. Another chance event is the number of accidents that will occur on a particular highway next week. Or, consider a machine that makes light bulbs. We know that a certain proportion of bulbs produced by that machine will be defective, hence every bulb made is a chance event as it could be defective. In this case, two possibilities exist, even if one of them (the production of a defective bulb) happens less frequently than the other. In short, chance events are at work in our daily lives as well as in the disciplines of science.

Which of the following are chance events?

A) The life span of a particular person.

B) A body of a gravity less than water floating in water.

C) Picking a card from a pack.

D) Expansion of a metal when heated.

E) In genetics, the prediction of the frequency of a particular bodily feature found under specified conditions.

F) Calculating a city's load of telephone calls at a given time.

In fact, A), C), E) and F) are events of probability.

A clarification

The reader may recall that we described mathematical logic as dealing with those propositions that could only be declared true or false and had no third possibility. Probability theory is somewhat different. It does not operate, as the term might imply, on chance events that can produce various results, but, rather, on the propositions that describe those chance events. Since these propositions involve uncertain outcomes, they are analyzed in terms of probability. While we seek truth-values in mathematical logic, in probability theory we seek probability-values. Of course, propositions concerning chance phenomena are themselves uncertain. Probability theory is a modern mathematical tool for analyzing propositions whose truth or falsehood is uncertain, or is certain only after the event. Consider this example: "In the next poker game I shall get a royal flush." Whether this is true or false can be decided only after the event; before the event we can merely say it has a certain probability.

Consider some further propositions about chance events: "The next throw of the die will be a six," (which has only a certain degree of probability); "The patient in bed fifteen will be in the hospital at most for one week," and so on.

For completeness, and for reasons which will become clear later, we include those propositions already known to be true, or false; that is, the limiting cases. The proposition, "The next throw of the die will produce a number not greater than 30," is necessarily true since the faces of a die are numbered 1 to 6. Such an event is called certain, and its probability value is 1. Denoting the probability by p and the above proposition by A, we have $p(A) = 1$, or the probability of A is 1, which means absolute certainty. The proposition, "The first number called in this Bingo game will be 100," is necessarily false because the numbers that can be called are 1 to 99. Denoting the proposition by B we have $p(B) = 0$, meaning the probability that B will occur is zero. Such an event is called impossible.

Hence, from information we have we can logically deduce that some events are certain, and some impossible. When neither of these limiting cases can be deduced the event is called possible, and is subject to probability assessments in the strictest sense. Probability theory, then, is a mathematical method for drawing acceptable conclusions concerning possible events.

Consider, finally, the following two arguments:

A) John is eighteen-years-old and yesterday he sprained his left ankle, therefore he will not run 100 meters in eleven seconds tomorrow.

B) All tigers are felines and all felines are mammals, therefore all tigers are mammals.

In B), the conclusion is implicit in the premises and has merely been made explicit by the argument; in A), the conclusion does not follow from the premises. No premise states that an eighteen-year-old with a sprained ankle cannot run 100 meters in eleven seconds. The conclusion tells us something that cannot be formally concluded from the premises. Of course, it is extremely likely that John will not be able to run 100 meters in eleven seconds with a sprained ankle, but it cannot be stated with absolute certainty. A margin of uncertainty remains, while none does in the second argument. We must therefore draw a distinction between those arguments in which the conclusion follows necessarily from the premises, and those in which it does not and, hence, can only be evaluated in terms of probability.

Probability theory, as the mathematical discipline that helps to check the arguments involving risk and uncertainty is, in this sense, a genuine logic. Therefore, we can say that logic, the science that checks the links between premises and conclusions, has two branches: deductive logic (including mathematical or formal logic) and inductive logic (probability).

fig. 229

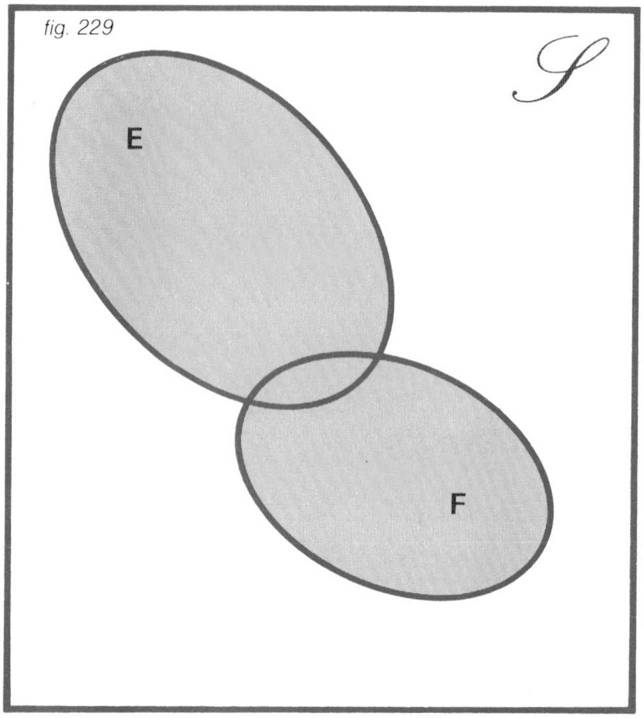

Sample space

Consider the throw of a coin, which can only fall heads or tails. Let us denote the set of outcomes by $S = \{h, t\}$. If we throw a die, one of the faces numbered 1 to 6 will come up; we denote the set of outcomes by $S = \{1, 2, 3, 4, 5, 6\}$. The set containing all possible outcomes of a chance event is called the sample space, while a particular outcome is the sample or sample point. An event is simply a set of outcomes, or rather, a subset of the sample space.

If we have a bowl containing six balls, one each of red (r), brown (b), violet (v), green (g), amber (a), navy-blue (n), we can represent the sample space by the set $S = \{r, b, v, g, a, n\}$. For clarity we shall confine ourselves to cases in which all outcomes can be expressed in finite space and time. There are mathematical methods for analyzing quite complicated events, which is clear if we examine slightly more complex cases.

Suppose our bowl contains two white and four red balls, the white ones being marked 5 and 6,

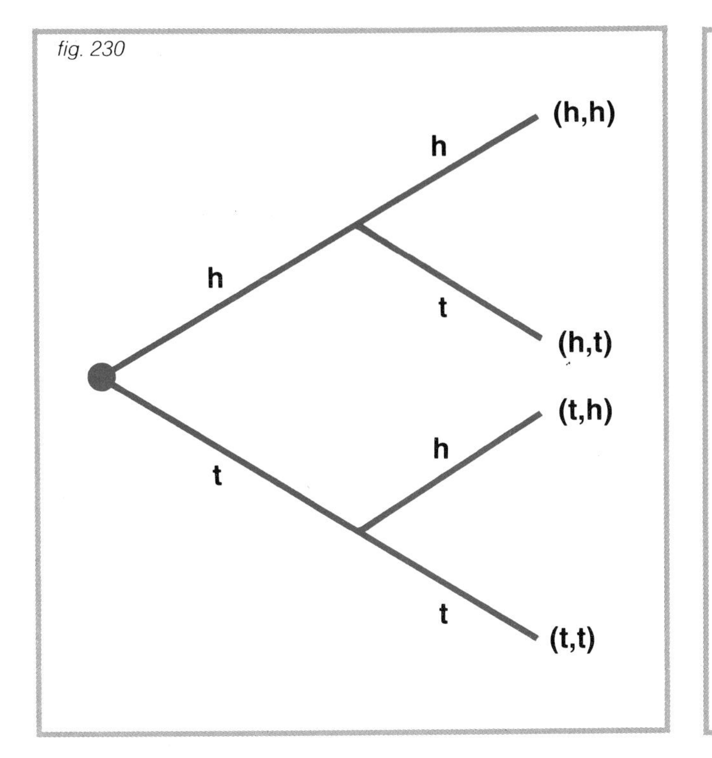

fig. 230

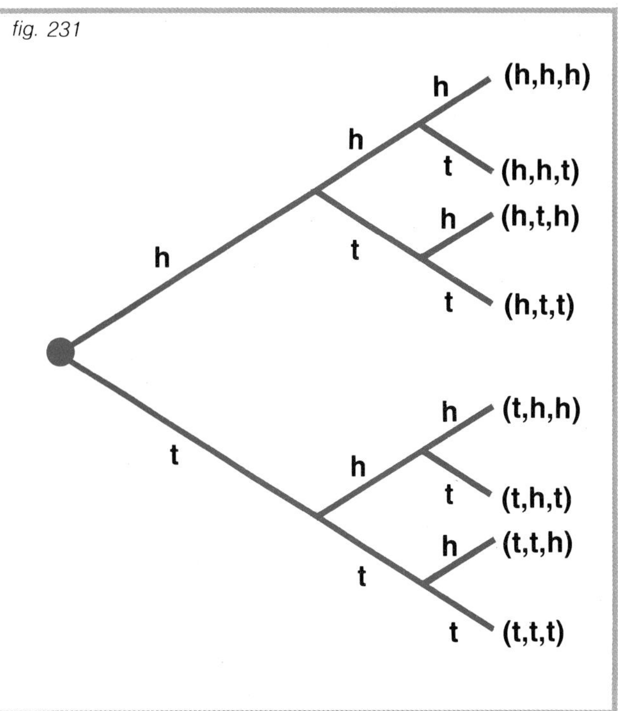

fig. 231

and the red balls from 1 to 4. The sample space for the chance event of extracting a ball is $S = \{1, 2, 3, 4, 5, 6\}$. Two events are possible in this situation: we extract a red ball or extract a white ball. If we call these two events E and F respectively, we can represent them as subsets of the sample space by writing $E = \{1, 2, 3, 4]$, $F = \{5, 6\}$. The sample space and the events can be represented by an Euler diagram. In Fig. 229, the rectangle is a sample space and the ellipses are the events E, F. In the present example, the empty set would correspond to an orange ball being extracted, or one marked 7, 8 and so on, all of which are impossible, so that $p(\varnothing) = 0$, or, translated into words: The probability of an impossible event is zero.

Consider another example: tossing two coins together. Let us write down the sample space, and the events of at least one coin falling tails up, and of both coins falling heads up. Thus $S = \{(h, h),\ (h, t),\ (t, h),\ (t, t)\}$, $E = \{(h, t),\ (t, h),\ (t, t)\}$, $F = \{(h, h)\}$. To determine sample spaces, it is often useful to draw tree diagrams that yield a picture of the elements of all events. The tree diagram for the present example is shown in Fig. 230, while Fig. 231 represents the case of three coins thrown in succession.

The measure of probability

Now, go back to the throw of a coin, where the sample space is $S = \{h, t\}$. If there is no reason to believe heads will appear more often than tails, the outcome h, or "heads", will occur with probability 1 in 2, ½ or 0.5. We write $p(h) = ½$. In general, we consider all the ways in which heads can come up and then divide by the number of elements in the sample space. This is intuitively obvious; the mathematical formulation makes the result general and allows us to analyze complex events. For a die the probability of the event, E, of throwing a six, is simply $p(E) =$ (number of ways for E to occur)/(number of elements in the sample space) $= 1/6$.

Suppose we have a bowl containing four balls, one black (b), one red (r), one white (w), and one amber (a). What is the probability that the next ball extracted will be red? The sample space is $S = \{b, r, w, a\}$ so that $p(r) = 1/4$. All probability values lie between 0 and 1. Thus we can write $0 \leq p \leq 1$; in words: The measure of probability p is a number greater than or equal to zero and less than or equal to one. If $p = 1$, the event is certain; if $p = 0$, the event is impossible. An example of a certain event is the sample space itself. In the case of throwing a coin we have $p(S) = p\{h, t\} = p(h) + p(t) = ½ + ½ = 1$. Here, the disjunction in "heads or tails" is translated by the sum of the probabilities (cf. p. 117).

Horse races

Let A, B, C be three horses in a race. One bettor claims that A's winning chances are twice B's, and B's are twice C's. If this is true, what are the various probabilities of winning? Let $p(A)$, $p(B)$, $p(C)$ be the respective probabilities for the horses, then $p(A) + p(B) + p(C) = 1$, since the sample space is a certain event. Moreover, we are told that $p(B) = 2p(C)$, $p(A) = 2p(B)$. Thus $p(A) = 4p(C)$ and $(4 + 2 + 1)p(C) = 1$. Therefore $p(C) = 1/7$, $p(B) = 2/7$, $p(A) = 4/7$.

The reader may have noticed that we made a tacit jump in logic, for it is not at all clear what is meant by saying that one contestant is twice as likely to win as another. In the case of drawing a ball from a bowl it is obvious what the sample space is; there are many different balls any one of which might be drawn. However, there are not, in this sense, several ways in which a race can be run. Of course, any one of the entries could win, but what does it mean to say one entry is more apt to win than another? What are the "different ways" to run the race, that correspond to the different ways to extract a white ball, say, from a bowl containing six white and four black balls? The two concepts of probability seem logically "incommensurable." What we do is this: We imagine a whole set of races, of which only one, in fact, will occur. If we now say that contestant A is twice as likely to win as contestant B, we mean that in the long run, A will win two out of every three races in the set, while B will win the remaining race. Notice the logical difference between the bowl containing balls and the race. In the former we can go on drawing balls but in the latter the event is finished once the race is run. However, by imagining the one race extended into a set of races, we can compare wins in a single race to draws from a bowl. Having grasped this point, the methods of calculation developed here can now be used.

The concept of function

We have used this concept implicitly; we must now define it explicitly. Take a common example: Every car has a registration, and for each car there is one, and only one, corresponding registration, and conversely. Hence, there is a one-one correspondence between the set of cars and the set of registrations, as shown in Fig. 232. An operation pairing each element of one set with one and only one element of another set if called a function. Each element in the left-hand column of Fig. 232 constitutes the "argument" of the function, while the right-hand column lists the "values" of the function (note the technical sense of the word "argument" as it is used here). In general we indicate functions by lower case letters *f,g,h....* For example, given any proposition A, the function relating A to its truth-values is $f(A) \in \{0, 1\}$.

When we discussed the truth-value function in propositional calculus, we specified that these values were attached to propositions concerning mathematical logic. Now we attribute a probability measure to a chance phenomenon. We do not actually assign the value to the phenomenon itself but rather to the proposition describing it. As the truth-value function operates on propositions, so the probability measure is a function that assigns a number between 0 and 1 to propositions relating to chance phenomena. Using *p* for the probability function, this is expressed by $p(\ldots) \in [0, 1]$,

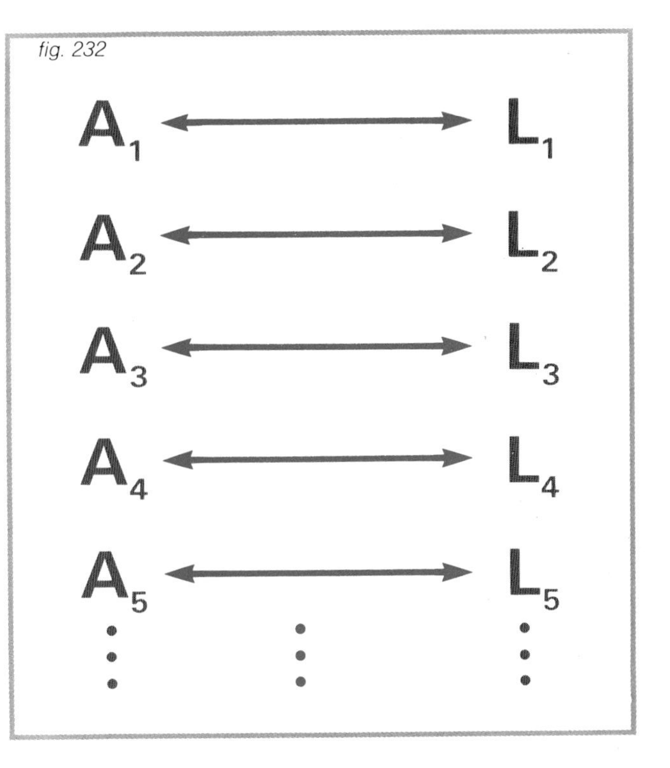

where the right-hand side indicates any number in the interval from 0 to 1.

The algebra of events and probability games

By combining probability theory and set theory we can clarify the analysis of complex problems. If events are sets, then any assertion about events can be translated into the language of sets, and conversely. Given a sample space *S* we can use set-theoretical operations to describe events in this space and to work out their probability measures.

The complementary event and its probability measure

For each event E there is an event $\bar{E}$, called its complement, which contains all outcomes in S but not in E.

Consider the chance phenomenon of throwing a coin, with $S = \{h, t\}$. The event of throwing heads is $E = \{h\}$, and the complement $\bar{E} = \{t\}$. Since $p(E) = ½$ we readily find $p(\bar{E}) = ½$. Using Venn diagrams we can illustrate the complement $\bar{E}$ as the area of S outside of E, as in Fig. 233.

In the present example the event and its complement have the same probability, however this is exceptional. Take the throw of a die and consider the event of six appearing. We have seen that $p(6) = ⅙$. The general formula for the probabilities of an event and its complement is $p(\bar{E}) = p(S) - p(E)$, as $\bar{E}$ contains all the elements of S outside E. Since $p(S) = 1$, we now have $p(\bar{E}) = 1 - ⅙ = ⅚$. The complement of the sample space S is obviously the impossible event $\varnothing$, and its probability is 0. Thus $\bar{S} = \varnothing$, and $p(\bar{S}) = 0$.

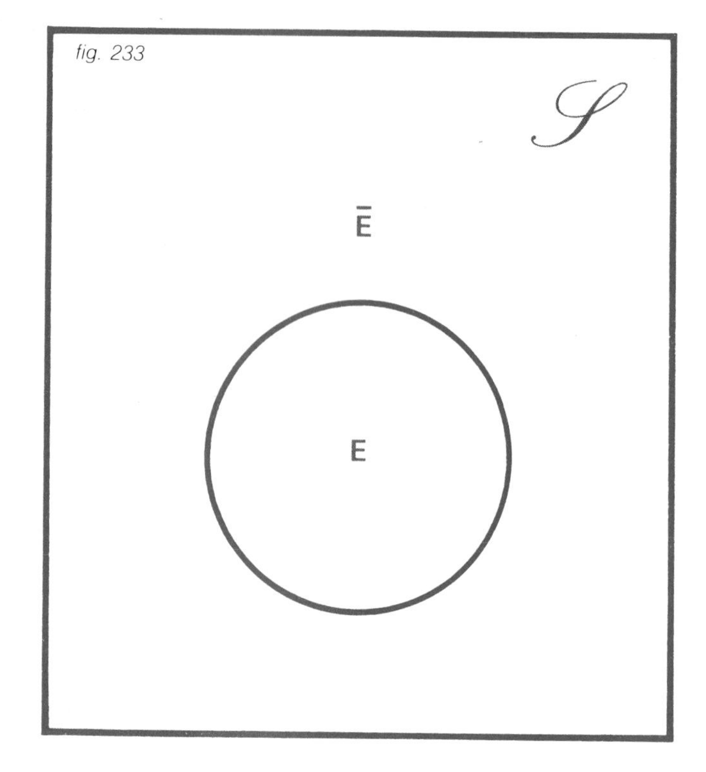

fig. 233

The probability of the union of two events

We have done this operation implicitly when referring to the probability of a certain event, such as the sample space. More generally, the union of the events E, F is $E \cup F$, with the event comprising all outcomes belonging to at least one of the two events. Thus $E \cup F$ occurs when E or F or both occur. Again take the throw of a die. The sample space is $S = \{1, 2, 3, 4, 5, 6\}$. Let G signify the event of the throw being even, H of the throw being odd, and I of the throw being prime. Thus

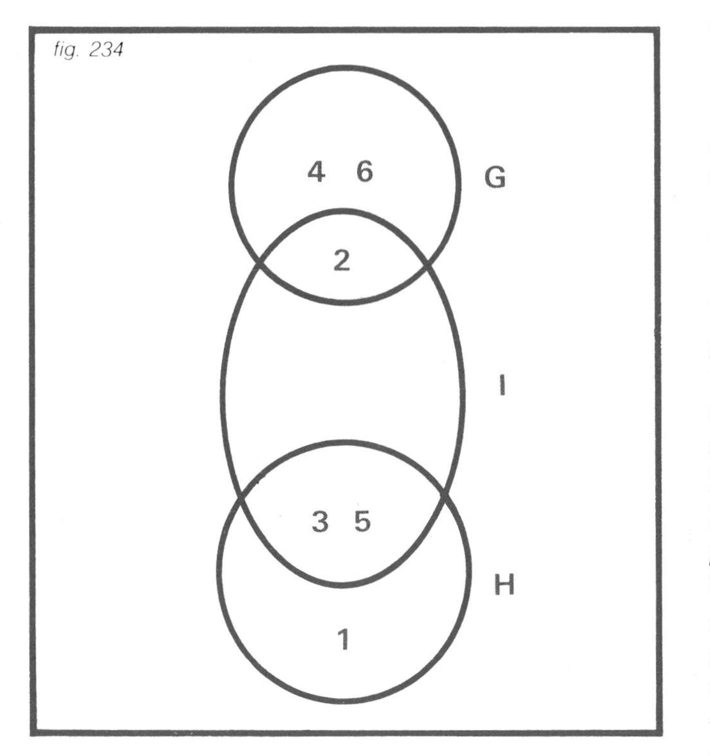

fig. 234

$G = \{2, 4, 6\}$, $H = \{1, 3, 5\}$, $I = \{2, 3, 5\}$. The event of the union of G and H is thus $G \cup H = \{1, 2, 3, 4, 5, 6\}$ which is S, while $H \cup I = \{1, 2, 3, 5\}$, and $G \cup I = \{2, 3, 4, 5, 6\}$. All this is shown in Fig. 234.

What is the probability of the union of G with H, that is $p(G \cup H)$? Since the union is equal to S, we have $p(G \cup H) = p(S) = 1$. This can be obtained less directly as follows. We know that $p(G \cup H) = p(G) + p(H)$. Now using our definition on p. 148, $p(G) = 3/6 = 1/2$ and $p(H) = 3/6 = 1/2$. Adding, we arrive at $p(G) + p(H) = 1/2 + 1/2 = 1$. That $G \cup H = S$ is, of course, peculiar to this example.

As we saw, $H \cup I = \{1, 2, 3, 5\}$. Let us therefore try, as before, $p(H \cup I) = p(H) + p(I)$. Since $p(I) = 3/6 = 1/2$, we would again have $p(H \cup I) = 1/2 + 1/2 = 1$, which is the probability of the entire sample space. Yet $H \cup I$ and S do not coincide, since $H \cup I = \{1,2,3,5\}$ and $S = \{1,2,3,4,5,6\}$, so their probabilities must be different. Obviously we have made a mistake, namely to suppose that $p(H \cup I) = p(H) + p(I)$. Going back to our definition of probability, $p(H \cup I) =$ (number of ways for $H \cup I$ to occur)/(number of elements in S). Now $H \cup I$ can occur in four ways, because it has four elements, while S has six. Therefore $p(H \cup I) = 4/6 = 2/3$, while $p(S) = 1$. Thus $p(H \cup I) \neq p(S)$. The source of our difficulty was in the fact that $H = \{1, 3, 5\}$ and $I = \{2, 3, 5\}$ have common elements; previously we had counted $p(3)$ and $p(5)$ twice.

The probability of the intersection of two events

The common elements of two sets are known as their intersection which is indicated by $\cap$ (cf. Fig. 156). In the present case, $H \cap I = \{3, 5\}$. This is shown diagrammatically in Fig. 235.

To evaluate $p(H \cap I)$, we divide the number of ways $H \cap I$ can occur by the number of elements in the sample space. Thus $p(H \cap I) = 2/6 = 1/3$. Now we can correct our mistake in calculating $p(H \cup I)$. When we add $p(H)$ to $p(I)$, we count the probability of the events {3} and {5} twice. We must therefore subtract the probability of the event {3, 5} which is exactly the intersection of the two sets. Thus quite generally, $p(H \cup I) = p(H) + p(I) - p(H \cap I)$, which in the present case yields $p(H \cup I) = 1/2 + 1/2 - 1/3 = 2/3$. The reason why we reached the correct answer when writing $p(G \cup H) = p(G) + p(H)$ is precisely because we have $p(G \cap H) = p(\varnothing) = 0$: The intersection of G and H in this instance has no members and is therefore the empty set (Fig. 236).

Note an important property of the empty set. Given an event E and its complement $\bar{E}$, their intersection consists of the impossible event, hence, $E \cap \bar{E} = \varnothing$. The two can never occur together. Such events are called disjoint or mutually exclusive. Thus the two events, "He is tall" and "He is short" cannot hold true for one person. Two compatible events that do not exclude each other would be, "He is tall" and "He is thin."

The probability of a choice

What we have learned so far helps us solve problems and puzzles in probability with greater clarity and logic.

Take a class of ten boys and twenty girls. Half the boys and half the girls have dark eyes. What is the probability that choosing one of them at random will yield a boy or a dark-eyed youngster? Let A signify the choice being a boy, and B the choice having dark eyes. The probability required is $p(A \cup B)$, which we know is equal to $p(A) + p(B) - p(A \cap B)$. The sample space $S = \{a_1, a_2, \ldots a_{29}, a_{30}\}$. Then $p(A) = 10/30 = 1/3$. $p(B)$ is

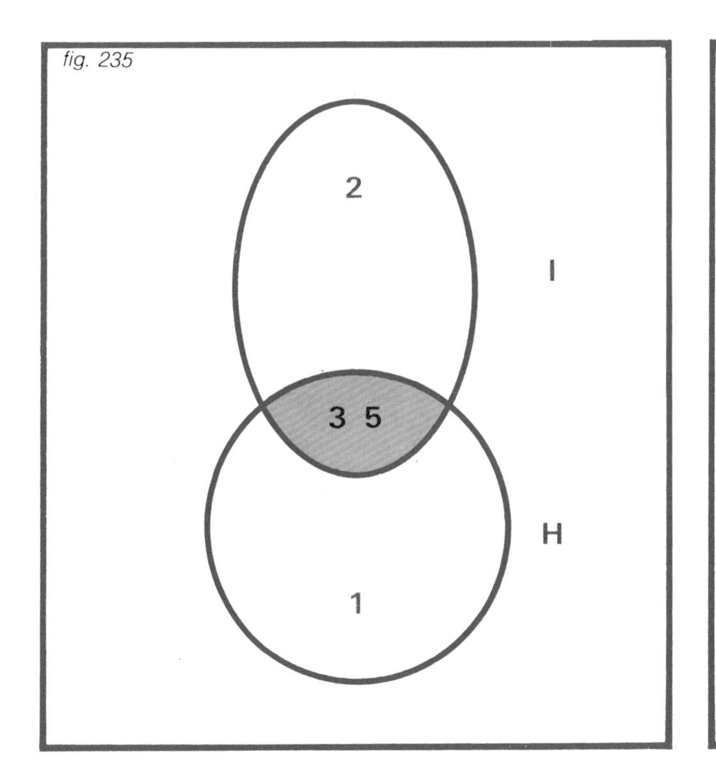

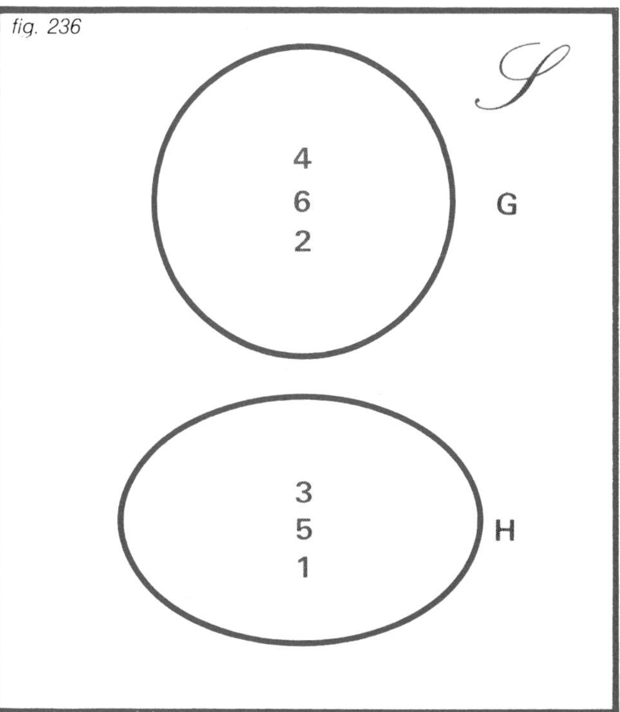

somewhat more complicated. Half the youngsters of each sex are dark-eyed, namely five boys and ten girls. Therefore $p(B) = 15/30 = 1/2$. There remains $p(A \cap B)$. Now the elements of $A \cap B$ are dark-eyed boys, of whom there are five, hence $p(A \cap B) = 5/30 = 1/6$. Now, we can finally calculate $p(A \cup B) = 1/3 + 1/2 - 1/6 = 4/6 = 2/3$.

Drawing a card from a pack

The familiar deck of 52 playing cards is a source of a variety of games based on probability. Consider the probability of drawing an ace or a king, in a single draw. To determine this we would do well to formalize the problem, and in that way reduce it to its essentials.

Let A be the event of drawing an ace, and K of drawing a king. Now we must find $p(A \cup K)$, which is equal to $p(A)+p(K)-p(A \cap K)$. The last term is obviously zero, for $A \cap K = \emptyset$: No single draw can be an ace *and* a king. As for $p(A)$ and $p(K)$: Since there are four aces and four kings in the pack, each of these terms is $4/52 = 1/13$. Therefore $p(A \cup K) = 2/13$.

Let H be the event of drawing a heart. What is the probability of drawing an ace or a heart? We have $p(A \cup H) = p(A) + p(H) - p(A \cap H)$. As seen before, $p(A) = 1/13$. Moreover, $p(H) = 13/52 = 1/4$, and $p(A \cap H) = 1/52$, since there is only one ace of hearts in the pack. Therefore $p(A \cup H) = 1/13 + 1/4 - 1/52 = 16/52 = 4/13$.

Joint throw of coin and die

In this instance, the sample space is given by $S = \{(h,1), (h,2), (h,3), (h,4)\ (h,5), (h,6), (t,1), (t,2), (t,3), (t,4), (t,5), (t,6)\}$. It has twelve elements. Each of the two ways in which the coin can fall is linked with the six possible falls of the die. This can be usefully represented, as it is in Fig. 237. Consider the compound event H: The coin falls heads up while the die shows an odd number. This is a subset of S and can occur in three ways: $H = \{(h,1), (h,3), (h,5)\}$, so that $p(H) = 3/12 = 1/4$, indicating that an arbitrary sample space can be a compound event. Indeed, the event "heads and three" is the intersection of "heads" for the coin and "three" for the die, and so too for all the elements of this sample space.

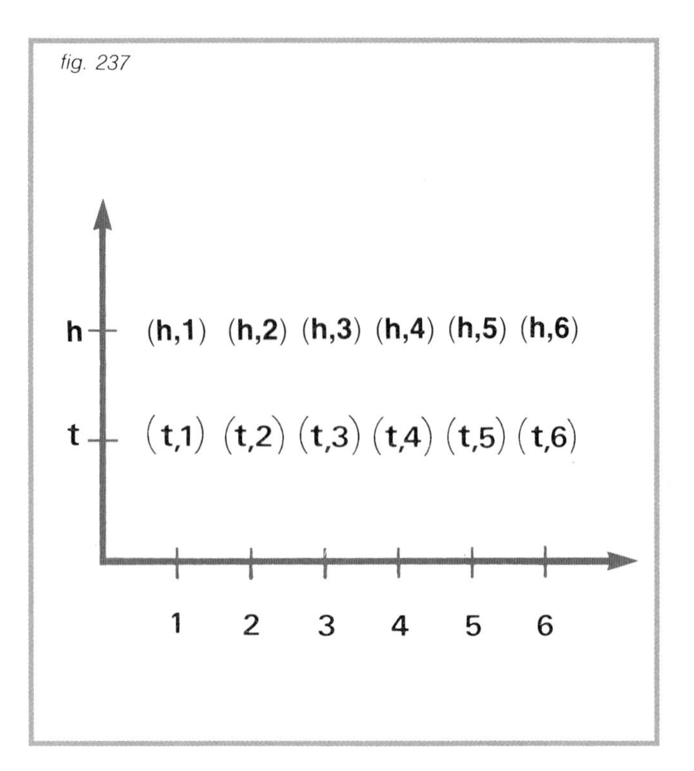

Dependent events

Suppose we have a bowl containing five balls, three white and two black. Let N_1 signify the first draw being black, and N_2 the second draw being black. Once the balls are drawn they are not put back in the bowl. What then is the probability that the second draw will be black if the first one is

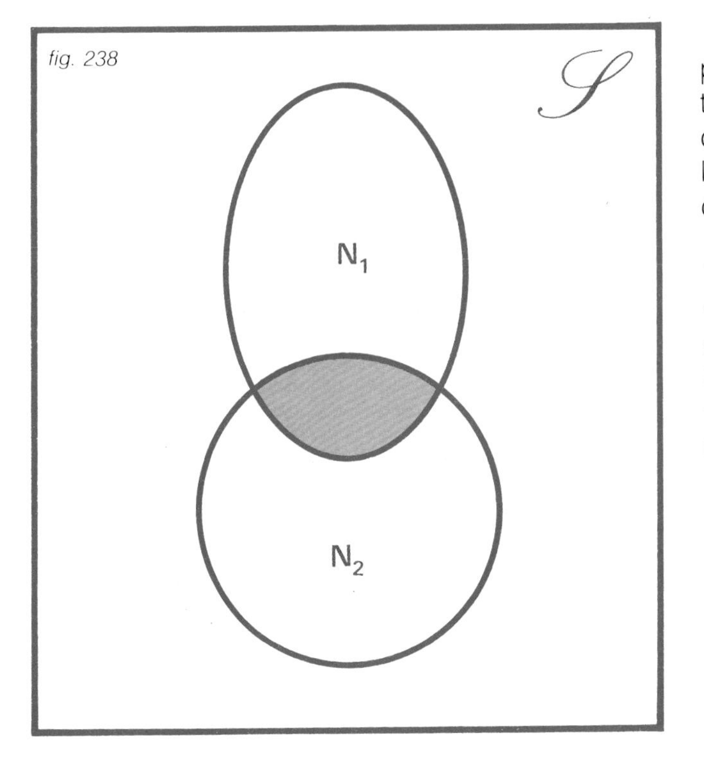

fig. 238

also? Here we need the concept of conditional probability, which is an extension, or refinement of previous concepts.

Given the two events N_1 and N_2, the conditional probability of N_2, given N_1 (the probability that N_2 will occur when N_1 has already occurred) is written as $p(N_2/N_1)$ and shown diagrammatically in Fig. 238.

Such situations occur in various fields. Suppose we must decide the following: A box contains 100 fragile crystal glasses of which 10 are defective. The first glass drawn is defective. How likely is it that the second glass will also be defective?

Let us go back to the bowl holding the balls. The sample space S is given by the five balls, and we want to evaluate $p(N_2/N_1)$. As N_1 has already occurred, the sample space is reduced to one black and three white balls, or a total of four balls, therefore $p(N_2/N_1) = 1/4$. This value depends on N_1 having occurred. If the first draw had been white, both black balls would remain in the bowl and the probability of N_2 would be $2/4 = 1/2$. This is not a different kind of probability, but rather a different kind of situation. In general, all probabilities are conditioned by certain assumptions about appropriate relations between the elements of the sample space. External events are never as simple as our account might suggest; even the simplest occurrence is interlocked in complex ways with other events or conditions, be they given or presupposed.

Thus, when we say the probability of throwing a six with a die is $1/6$, we should really say $p(6/\text{true die}) = 1/6$, or, in words: The probability of throwing a six, assuming the die to be unbiased, or "true", is $1/6$. Now the reader can solve the problem of the crystal glasses.*

*The probability of the second draw being defective is $1/11$.

Independent events

Events are called independent if the occurrence of one does not affect the probability that the second will occur in turn. Note that if the occurrence of the two events A and B is independent, then $p(B/A) = p(B)$; the fact that A has taken place does not bear on the probability that B will take place. Similarly $p(A/B) = p(A)$.

We can now define independent events formally. Since generally $p(B/A) = p(A \cap B)/p(A)$, we have $p(B) = p(A \cap B)/p(A)$, or $p(A \cap B) = p(A)p(B)$. In words: Two events A and B are independent if the probability of their joint occurrence is equal to the product of the probabilities of each one. These notions are readily applied to actual cases.

Now back to the joint throw of coin and die. Each element was the intersection of two events. The probability of each was $1/12$, as there were 12 elements in all. For example, $p(h, 1) = 1/12$; this is the intersection of first, a coin being thrown and next, a die being thrown. The probability then of throwing heads and a 1 are, respectively $p(h) = 1/2$ and $p(1) = 1/6$. Now $p(h)p(1) = 1/2 \cdot 1/6 = 1/12 = p(h, 1) = p(h \cap 1)$, hence the events are independent.

The same is true in the example seen earlier of a joint throw of a coin and a die. If D is the event of throwing an odd number, then $p(h \cap D) = p(h, D) = p(H) = 1/4$, as on p. 154. Now $p(h) = 1/2$ and $p(D) = 1/2$, since half the numbers on the die are odd. Thus $p(h)p(D) = 1/2 \cdot 1/2 = 1/4$, yielding $p(H \cap D) = p(h)p(D)$ and revealing that h and D are independent.

Probability calculus originated with games of chance—dice, cards and bets. Such games were recorded as early as 2000 years ago; even in the Greco-Roman world there were professional gamblers. Probability is generally understood as a frequency: the number f of times an event E occurs in n possible cases. If each occurrence is independent of the others they are equi-probable, and $p(E) = f/n$, the ratio of occurrences to total cases. Theoretically a true coin *(right)* could fall heads down every time it is thrown, but actually it will fall heads up roughly half the time. The ratio of heads to the total number of throws is the relative "frequency" of heads. This kind of probability, sometimes called statistical because frequency is a statistical concept, is widely used and particularly in games of chance in which observations can be repeated under very similar circumstances. This is not the only interpretation of the mathematical concept of probability. At least one other, known as "subjective," measures the degree of confidence a person places in an event occurring, given his opinions and information.

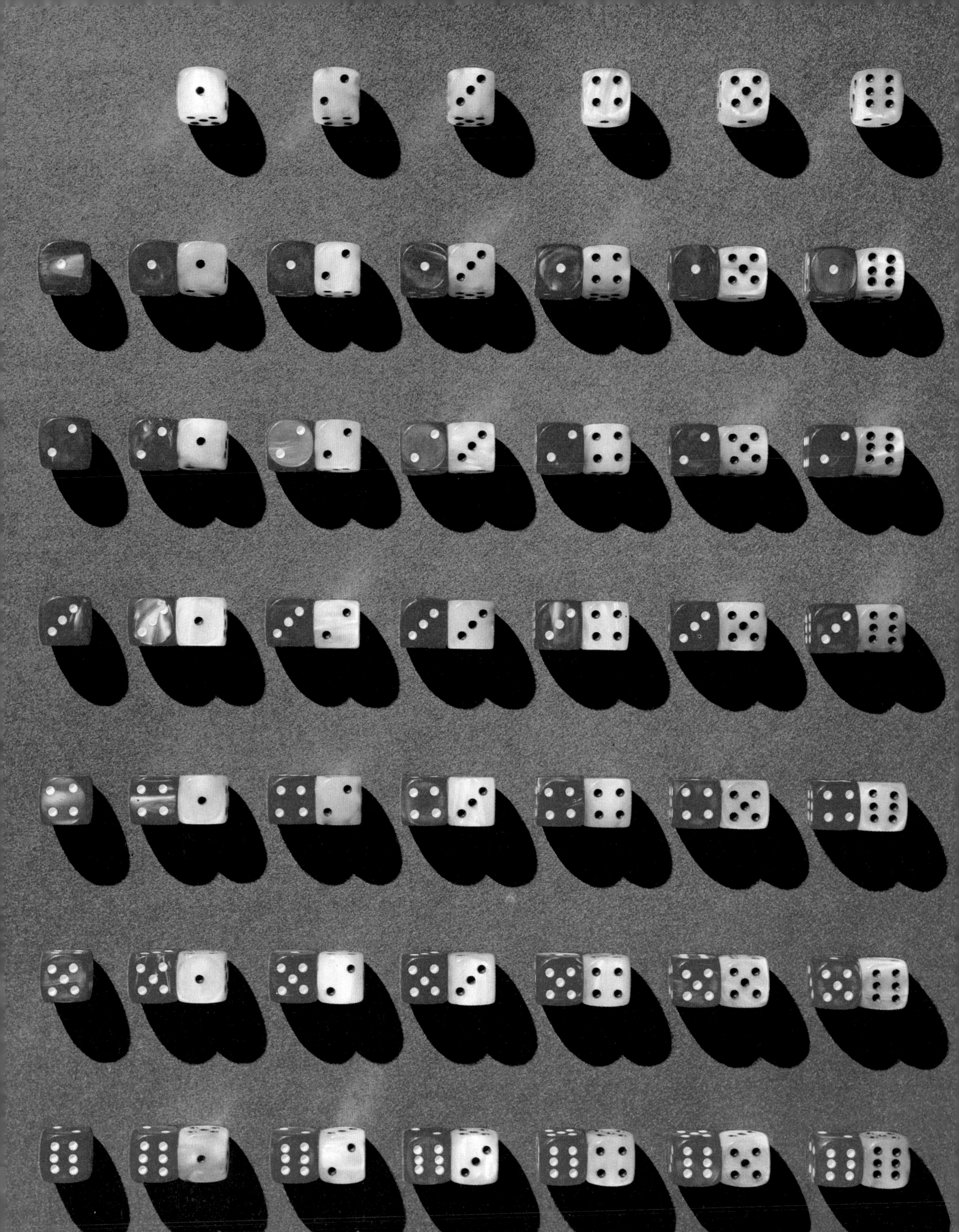

The throw of a die can be interpreted in terms of frequency. Suppose we are interested in the throw of a number less than four. If after ten throws a number less than four appears four times, the relative frequency is $4/10$ or 0.4. If we take the compound event of throwing two dice, there are 36 cases *(left)*. If we want a throw whose total is seven, there are six possibilities: (1, 6), (2, 5), (3, 4), (4, 3), (5, 2), (6, 1) and the probability of the event is $6/36 = 1/6$. In recent years the probability theory has grown in importance in almost every area of our lives. Such games of chance as football pools and lotteries have made most of us aware of the notion of probability.

Many of today's problems—traffic density and control, for example—are attacked by coupling probability with the theory of graphs. The results of medical research are expressed in terms of probability, and in nuclear physics certain laws expressed by probability functions are providing clearer research guidelines.

What is the probability that George and Bob speak the truth?

There are, of course, some problems involving probabilities that cannot be resolved by intuition alone.

George and Bob are in the habit of telling fibs. George tells the truth three times out of four and Bob four times out of five. When they both say the same thing, what is the probability that the statement is true? If G is the event of George making an assertion, which we know may be true or false, $G = \{T_g, F_g\}$; the T_g stands for George's assertion being true, the F_g for false. Similarly we write $B = \{T_b, F_b\}$. Further, we are given $p(T_g) = 3/4$, $p(T_b) = 4/5$, making $p(F_g) = 1/4$, $p(F_b) = 1/5$. These events are not equally probable, but nevertheless they are independent. The sample space for G and B making some assertion is $S = \{(T_g, T_b), (T_g, F_b), (F_g, T_b), (F_g, F_b)\}$. Also $p(T_g, T_b) = p(T_g \cap T_b) = p(T_g)p(T_b) = 3/4 \cdot 4/5 = 3/5$. Similarly $p(T_g, F_b) = 3/20$, $p(F_g, T_b) = 1/5$, $p(F_g, F_b) = 1/20$. Clearly, $p(S) = 3/5 + 3/20 + 1/5 + 1/20 = 1$.

Now let C be the event of George and Bob making the same statement; that is, C is the subset $\{(T_g, T_b), (F_g, F_b)\}$. Both speak the truth or both lie.

This event can occur in two ways, so that $p(C) = p(T_g, T_b) + p(F_g, F_b) = 3/5 + 1/20 = 13/20$ is the probability that both make the same statement. Finally, we must determine the probability that when George and Bob make the same statement they are speaking the truth. This is clearly a conditional probability. By applying the general formula to this case, $p((T_g, T_b)/C) = p((T_g, T_b) \cap C)/p(C) =$ (probability of both speaking the truth and making the same statement)/$(13/20) = (12/20)/(13/20) = 12/13$.

Probability and empirical science

The concepts of dependent and independent events give us a mathematical model by which we can analyze the relations between a variety of events. Hence, we can decide whether such relations exist, and if so, to what extent. However, we must be certain to distinguish between statistical independence, which is what we have been examining, and the independence between physical phenomena. When we say the occurrence of event *A* affects or changes the probability of event *B*, we do not mean *A* is the cause of *B* in the physical or material sense. The kind of dependence or independence discussed here is only statistical; it helps us to describe and mathematicize certain relations between facts. The interpretation of those relations, however, is a task for the researcher and constitutes a conceptual leap based on many factors, not the least of which is the scientific theory guiding the investigation. Why, for instance, are doctors looking for carcinogenic agents in tobacco? Because there is an incontrovertible statistical link between smoking and lung cancer, making a physical link between them also probable. In short, probability is a unique tool for analyzing and guiding research in all scientific disciplines.

Probability and statistics

In large measure, scientific knowledge progresses because of the concrete applications of probability. A doctor wishing to check the effectiveness of a particular drug, carries out a series of experiments; similarly an agronomist, wishing to determine whether a particular fertilizer will improve the yield of a crop given certain constants—amount of sunlight, type of soil, irrigation—performs tests on the field. The use of objective techniques to test the results of scientific experiments allows scientists to draw conclusions and move from the specific to the more general.

The agronomist, for instance, might extend his conclusions to all plants of the same species as those he examined, while the doctor might use his conclusions to help all patients suffering from a disease similar to the one affecting the test patients.

This course from the specific to the general follows several steps. An individual case is examined, and a limited number of observations are made on the basis of which conclusions about all such cases are drawn. Some thinkers, past and present, deny the procedure's legitimacy and reject any general conclusion drawn from a particular observation. And, of course, it is true that such conclusions do involve a degree of uncertainty.

It is at this point that probability is useful, for the uncertainty can be measured by a probability-value. Hence the probability approach is important to our knowledge of reality as most of the assertions we make about reality are only probably true, that is true within certain limits which can be specified by probability-values. Statistics gives us the logical and mathematical techniques to make inferences and then measure their degree of uncertainty.

Sample and population

In statistics, the term "population" is a technical one and refers to the range of occurrence of the characteristic being examined. If we are seeking a relation between the height and weight of 18-year-old students in a given city, the population consists of all students in that city. If we need to know the efficiency of a certain fertilizer, the population consists of all comparable fertilizers. Then too, a population could be an infinite number of throws of a coin from which we want to determine the number of times heads appear.

Although, generally, the aim of experimental research is to acquire knowledge about an entire population, the size of the population frequently makes it impossible to do so. In that event we confine the examination to a sample. Conclusions drawn from samples and applied to populations are called statistical inferences. In some measure these inferences are uncertain, and for the degree of uncertainty to be measurable in terms of probability, the sample must be chosen at random.

Guess the vintage

Suppose a wine connoisseur claims he can not only distinguish between the quality of various wines but also between their years of production. Let us find out if his information is reliable or merely accidental. Blindfolding the expert, we present him with several wines from different years. Either he is simply a braggart and will give the right answer by chance, or he has a sensitive palate and can make the correct distinctions. In the case of the former his answers will be random and just as apt to be right as wrong. If G is the event of a correct answer, and H the event of a wrong one, the sample space is $S = \{G, H\}$, and $p(G) = p(H) = 1/2$. If the first answer is correct, the expert is given a second test; the sample space is now $S = \{(G, G), (G, H), (H, G), (H, H)\}$. If the answers were (G, G), we try him again, with the new sample space of $S = \{(G, G, G), (G, G, H), (G, H, G), (G, H, H), (H, G, G), (H, G, H), (H, H, G), (H, H, H)\}$. Should the answer be correct a third time, the probability of it being chance is one-eighth. The more the degree of uncertainty declines, the more confident we are that the expert is indeed an expert, even if a margin of uncertainty does remain.

Clearly one correct answer is not convincing, as the chance of being right is one out of two. However, the more successive answers are correct the less likely it is that it is due to chance. Statistical inference reveals the limitations within which our assertions about the world are correct.

Conclusion

The concepts we have outlined are the basis of the modern mathematical theory of probability. Linked to these concepts is the theory of statistics, essential to researchers in empirical science and widely applied in our daily lives. For further study, see bibliography (p. 179).

APPENDIX: GAMES WITH LOGIC AND PROBABILITY

Note

New mathematical concepts can appear abstract and theoretical and the reader might well consider the final two chapters out of tune with the book's title. However, the basic concepts of mathematics, logic and probability should be more familiar, which is why we presented an elementary account of propositional calculus and several introductory concepts of probability theory. In doing so we deliberately refrained from presenting a number of games and concentrated instead on explanations of the concepts; solving even elementary problems and games with logic and probability would be far more difficult were we to trust intuition alone. Therefore, we offer the following appendix of games, problems and exercises as compensation for the theoretical topics covered in the last chapters, and we assume that while working the problems the reader will understand the reason for our theoretical explanations. Although the problems become increasingly complex, they do not require information that has not been covered. We begin with logic and move on to probability.

Games with logic

1.1 - Can the following be formulated in propositional logic? Why?

a) Charles and John are athletes.

b) There is probably life on Mars.

c) If the water is clear, then George can see the riverbed.

d) John thought Mary had been delayed by engine trouble.

e) The fire was caused by a shortcircuit or by combustion.

f) Mary and John are married.

g) If there are more cats than dogs, then there are more horses than dogs and fewer snakes than cats.

h) Dick and Tom like drinking.

Solution.

a) Yes, because this can be interpreted as the conjunction of two propositions: "Charles is an athlete and John is an athlete."

b) No. This concerns probability and is neither true nor false.

c) Yes. This is an implication.

d) No, because there is no point in asking whether "John thought X" is true or false.

e) Yes. This is a disjunction.

f) No. This expresses a relation, and it cannot be interpreted as a conjunction since the proposition, "Mary and John are married" is not equivalent to, "Mary is married and John is married."

g) Yes. This is an implication.

h) Yes. This is a conjunction.

1.2 - Express in propositional form those statements in 1.1 that can be so formulated in propositional logic.

Solution.

a) $C \wedge J$, *with* C = *Charles is an athlete*, J = *John is an athlete.*

c) $L \rightarrow F$, *with* L = *the water is clear*, F = *George can see the riverbed.*

e) $A \vee B$, *with* A = *the fire was caused by a shortcircuit*, B = *the fire was caused by combustion.*

g) $S \rightarrow (H \wedge G)$, *with* S = *there are more cats than dogs*, H = *there are more horses than dogs*, G = *there are fewer snakes than cats.*

h) $D_1 \wedge D_2$, *with* D_1 = *Dick likes drinking*, D_2 = *Tom likes drinking.*

1.3 - Determine the truth-values of the compound propositions a, b, c, given the truth-values of the following premises.
W = "Galileo was born before Newton," true;
J = "Newton was born in the 17th century," true;
K = "Newton was born before Fermat," false;
L = "Leibniz was a compatriot of Galileo," false.

a) If Galileo was born before Newton, then Newton was born before Fermat.

b) If Leibniz was a compatriot of Galileo or Newton was born before Fermat, then Newton was born in the 17th century.

c) If Leibniz was not a compatriot of Galileo, then Newton was born before Fermat or he was not born in the 17th century.

Solution.

	$a = W \rightarrow K$;	$b = (L \vee K) \rightarrow J$;	$c = \neg L \rightarrow (K \vee \neg J)$
truth-values	$1\ \underline{0}\ 0$	$0\,0\,0\ \ \underline{1}\ 1$	$1\,0\ \underline{0}\ \ 0\,0\,0\,1$

Thus a is false, b is true, c is false.

1.4 - If P, Q are true and R false, what are the truth-values of:

a) $\neg P$;

b) $\neg(P \wedge R)$;

c) $\neg(P \vee Q)$;

d) $P \vee (Q \wedge R)$;

e) $R \rightarrow ((Q \wedge R) \vee (P \vee Q))$;

f) $R \leftrightarrow (P \wedge R)$?

Solution.

a) $\neg P$; b) $\neg(P \wedge R)$; c) $\neg(P \vee Q)$; d) $P \vee (Q \wedge R)$

01   1 1 0 0   0 1 1 1   1 1 1 0 0

e) $R \rightarrow ((Q \wedge R) \vee (P \vee Q))$ f) $R \leftrightarrow (P \wedge R)$

0 1   1 0 0 1 1 1 1   0 1 1 0 0

1.5 - Determine whether the following inference is logically valid or not (cf. p. 124–125): The train is defective or the current is off. The current is not off. Therefore the train is defective.

Solution.
Let D = the train is defective, E = the current is off, and construct the following truth-table:

Premises					Conclusion
D	∨	E	¬	E	D
1	1	1	0	1	1
0	1	1	0	1	0
1	1	0	1	0	1
0	0	0	1	0	0

In the only case where both premises are true (line 3), the conclusion is also true, hence the argument is valid. Note that with the conclusion ¬D, the argument is invalid because the truth-table has a line of true premises with a false conclusion:

Premises					Conclusion	
D	∨	E	¬	E	¬	D
1	1	1	0	1	0	1
0	1	1	0	1	1	0
1	1	0	1	0	0	1
0	0	0	1	0	1	0

1.6 - Prove the validity of the following argument: If we do not increase the number of jobs in our society, crime will rise. Crime is not rising in our society. Therefore the number of jobs is increasing.

Let W = the number of jobs is increasing, D = crime is rising.

Solution.

Premises					Conclusion
W	∨	D	¬	D	W
1	1	1	0	1	1
1	1	0	1	0	1
0	1	1	0	1	0
0	0	0	1	0	0

1.7 - Determine whether the following argument is logically valid: John did not receive our note or he has made other arrangements. John did receive our note. Therefore he has not made other arrangements.

Let P = John did not receive our note, R = John has made other arrangements.

Solution.

Premises					Conclusion	
P	∨	R	¬	P	¬	R
1	1	1	0	1	0	1
0	1	1	1	0	0	1
1	1	0	0	1	1	0
0	0	0	1	0	1	0

The argument is invalid.

1.8 - Show that the following argument is valid: If the budget is not cut, prices will remain stable if, and only if, taxes are raised. Taxes will be raised only if the budget is not cut. If prices remain stable, taxes will not be raised. Therefore taxes will not be raised.

Let B = the budget is cut, P = prices will remain stable, R = taxes will go up.

Solution.

	Premises		*Conclusion*
$\neg B \rightarrow (P \leftrightarrow R)$	$R \rightarrow \neg B$	$P \rightarrow \neg R$	$\neg R$
0 1 1 1 1 1	1 0 0 1	1 0 0 1	0 1
1 0 1 1 1 1	1 1 1 0	1 0 0 1	0 1
0 1 1 0 0 1	1 0 0 1	0 1 0 1	0 1
1 0 0 0 0 1	1 1 1 0	0 1 0 1	0 1
0 1 1 1 0 0	0 1 0 1	1 1 1 0	1 0
1 0 0 1 0 0	0 1 1 0	1 1 1 0	1 0
0 1 1 0 1 0	0 1 0 1	0 1 1 0	1 0
1 0 1 0 1 0	0 1 1 0	0 1 1 0	1 0

The argument is valid because in all three cases in which all the premises are true (lines 5, 7, 8), the conclusion is also true.

1.9 - Test the following argument for validity: If, in a chemical experiment, an orange mixture is formed, sodium or potassium is present. If there is no sodium there is iron. If there is iron and an orange mixture is formed there is no potassium. Therefore sodium is present.

Let A = an orange mixture is formed, S = sodium is present, P = potassium is present, F = iron is present.

Solution.
As before, we could draw a complete truth-table with four variables, listing all sixteen possibilities. The table would then prove the argument invalid since there are cases in which all the premises are true but the conclusion is false. It is enough, however, to show that at least one such case occurs (take A *and* S *as false,* P *and* F *as true):*

$A \rightarrow (S \vee P)$	$\neg S \rightarrow F$	$(F \wedge A) \rightarrow \neg P$	S
0 1 0 1 1	1 0 1 1	1 0 0 1 0 1	0

1.10 - Represent the following statements with Venn diagrams:

a) Some cats are wild.

b) No bird wears a vest.

c) Some dogs are white.

d) Some men are five feet tall, and all these are jockeys.

Solution.

a) G = {*cats*}, S = {*wild animals*}

b) U = {*birds*}, E = {*vest wearers*}

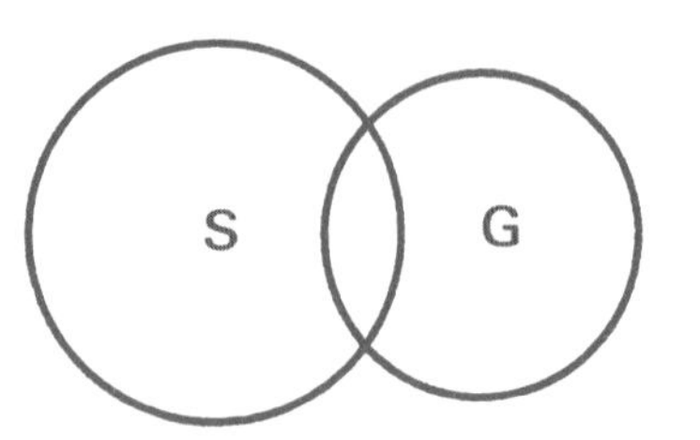

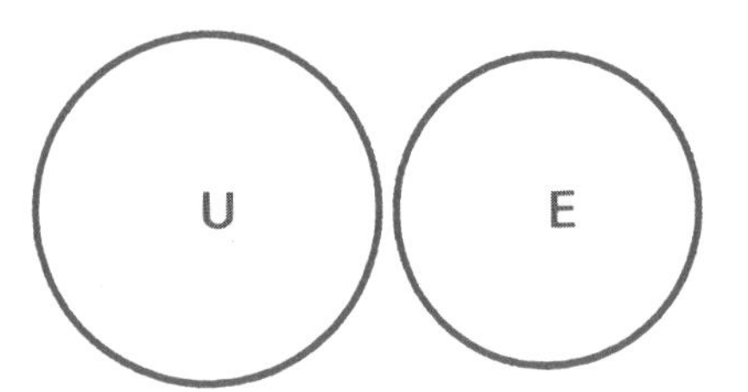

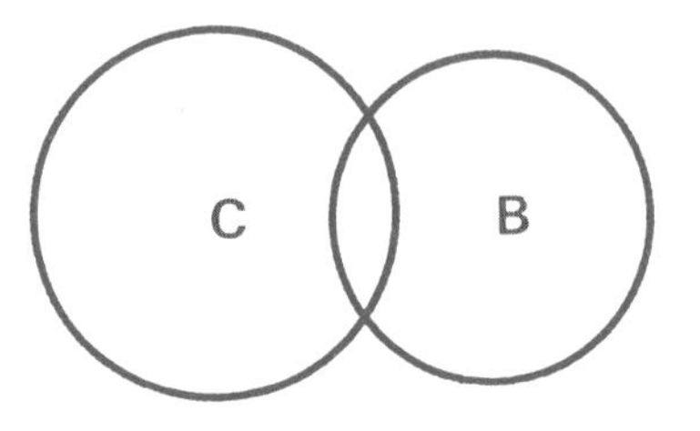

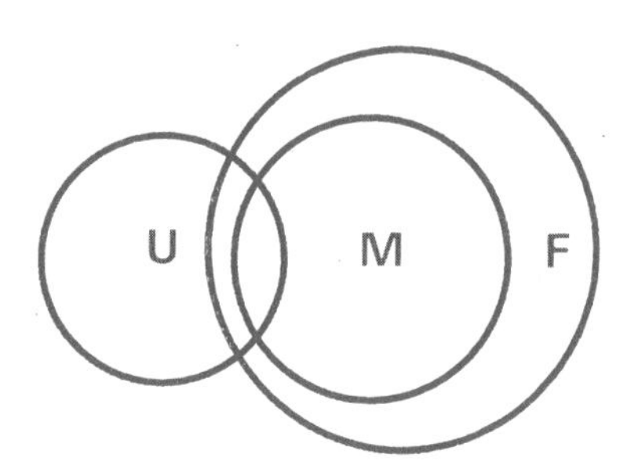

c) C = {*dogs*}, B = {*white things*}

d) U = {*men*}, M = {*things five feet tall*}, F = {*jockeys*}

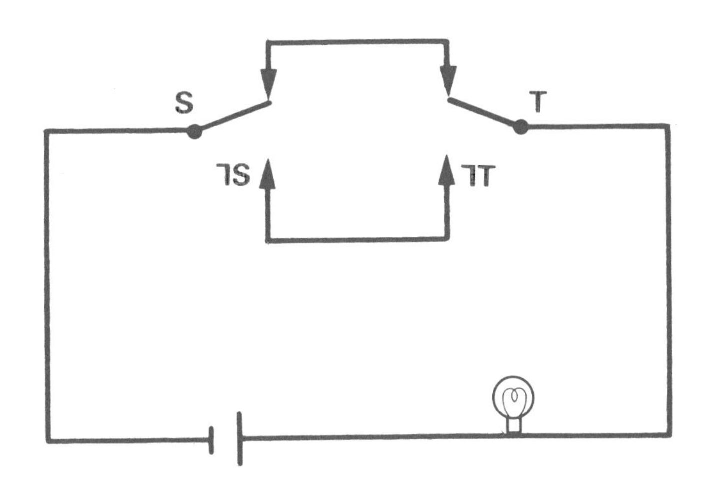

1.11 - Use Venn diagrams to test the validity of the following argument: If all pupils are insane and all criminals are insane, then all pupils are criminals.

Use the following sets: *A* = {pupils}, *B* = {insane people}, *C* = {criminals}.

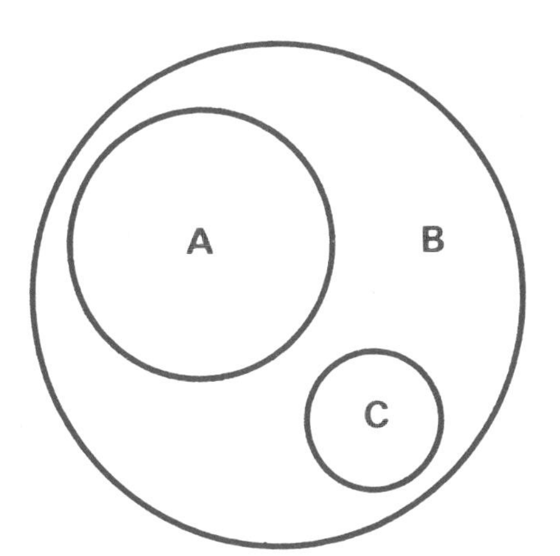

Solution.
Without reviewing all the possible combinations, the following diagram proves the argument is invalid.

1.12 - John and Mary live in a large apartment with a long corridor; the light can be switched on only at one end of the corridor. What electric circuit will let them control the light from either end of the hall, and what is its algebraic expression (cf. pp. 135–141)?

Solution.
The switches can be on or off (with the lever up or down respectively). We need the kind of circuit that ensures the light is on when both switches are up and also when both switches are down. Given the various truth-tables, this clearly leads to the biconditional. Indeed $A \leftrightarrow B$ *is equivalent to* $(A \rightarrow B) \wedge (B \rightarrow A)$. *(cf. p. 124).*

Games with probability

2.1 - Two coins are tossed at the same time with equal probability that each will fall on one side or the other. What is the probability of:

a) two heads,

b) one heads and one tails?

Solution.
The sample space is S = {(h, h), (h, t), (t, h), (t, t)} *hence*

a) p(h, h) = $\frac{1}{4}$, *and*

b) p((h, t), (t, h)) = $\frac{1}{4}+\frac{1}{4}=\frac{1}{2}$.

2.2 - Three coins are tossed at the same time. What is the probability of the event *A* = at least one heads, and *B* = all heads or all tails?

Solution.
The sample space is $S = \{(t, t, t), (t, t, h), (t, h, t), (h, t, t), (h, h, t), (h, t, h), (t, h, h), (h, h, h)\}$. *We might also write this as* $\{0, 1, 2, 3\}$, *referring to the number of heads. Thus* $p(0) = 1/8$, $p(1) = p(2) = 3/8$, $p(3) = 1/8$.
Then $A = \{1, 2, 3\}$, *so that* $p(A) = p(1) + p(2) + p(3) = 3/8 + 3/8 + 1/8 = 7/8$; *and* $B = \{0, 3\}$, *so that* $p(B) = p(0) + p(3) = 1/8 + 1/8 = 1/4$.

2.3 - In a defective coin where heads are twice as probable as tails, what is the probability $p(h)$ and $p(t)$?

Solution.
We are given $p(h) = 2p(t)$. *Now* $S = \{h, t\}$, *and* $p(S) = p(h) + p(t) = 1$, *thus* $3p(t) = 1$, $p(t) = 1/3$, $p(h) = 2/3$.

2.4 - Colin, Mark and Nick take part in a race. Colin and Mark are equally apt to win, and each is twice as apt to win as Nick. What is the probability that either Mark or Nick will win?

Solution.
Let C = *Colin wins*, M = *Mark wins*, N = *Nick wins. We have* $p(C) = p(M) = 2p(N)$ *and we require* $p(M \cup N) = p(M) + p(N)$. *Now* $p(S) = 2p(N) + 2p(N) + p(N) = 5p(N) = 1$. *Hence* $p(N) = 1/5$. *Therefore* $p(C) = p(M) = 2/5$ *and* $p(M \cup N) = 3/5$.

2.5 - Without looking, we draw a ball from a bowl containing 15 balls numbered 1 to 15. If each is just as apt to be drawn as the others, what is the probability that the number on the ball is:

a) divisible by three,

b) even,

c) odd,

d) a perfect square?

Solution.
The sample space is $S = \{1, 2, 3 \ldots 13, 14, 15\}$. *Divisibility by three is the event* $A = \{3, 6, 9, 12, 15\}$, *an even draw* $B = \{2, 4, 6, 8, 10, 12, 14\}$, *an odd draw* $C = \{1, 3, 5, 7, 9, 11, 13, 15\}$, *a perfect square draw* $D = \{1, 4, 9\}$. *Then:*

a) $p(A) = 5/15 = 1/3$;

b) $p(B) = 7/15$;

c) $p(C) = 8/15$;

d) $p(D) = 3/15 = 1/5$.

2.6 - Given a box with 6 red, 4 white and 5 blue balls, one ball is drawn without looking. What is the probability that it will be:

a) red,

b) white,

c) blue,

d) not red,

e) red or white?

Solution.
The sample space is S = {6 r, 4 w, 5 b}. *Let* R = *a red draw,* W = *a white draw,* B = *a blue draw.*

a) $p(R) = 6/15 = 2/5$;

b) $p(W) = 4/15$;

c) $p(B) = 5/15 = 1/3$;

d) $p(\overline{R}) = 1 - p(R) = 1 - 2/5 = 3/5$;

e) $p(R \cup W) = p(R) + p(W) = 2/5 + 4/15 = 10/15 = 2/3$; *alternatively,*
$p(R \cup W) = p(\overline{B}) = 1 - p(B) = 1 - 1/3 = 2/3$.

The formula used for the union is the same as the one for mutually exclusive events (cf. p. 151).

2.7 - A bowl contains 200 balls numbered 1 to 200, all equally apt to be drawn. What is the probability of drawing a ball with a mark divisible by 6 or 9?

Solution.
The sample space is S = {1, 2 . . . 199, 200}. *Between 1 and 200 there are* [200/6] *numbers divisible by 6, the square bracket indicating the biggest integer in the quotient. Thus* [200/6] = 33. *Similarly,* [200/9] = 22. *Now every third integer divisible by 6 is also divisible by 9, so we must subtract* 33/3 = 11. *Hence the probability is* 44/200 = 11/50.

2.8 - If John has three chances out of ten of winning a certain game, how likely is he to lose?

Solution.
The losses are complementary to the gains, hence the likelihood of his losing is $1 - 3/10 = 7/10$.

2.9 - A true die is thrown. What is the probability of throwing *Q*: a 4; *I*: less than 4; *R*: an even number; *T*: an odd number?

Solution.
The sample space is S = {1, 2, 3, 4, 5, 6} *and* $p(Q) = p(4) = 1/6$; $p(I) = p(1) + p(2) + p(3) = 3/6 = 1/2$; $p(R) = p(2) + p(4) + p(6) = 1/2$; $p(T) = p(1) + p(3) + p(5) = 1/2$.

2.10 - Tom and Jack each throw a die. If Tom's throw is the higher, he gives Jack a token and conversely; if the throws are equal, no token is exchanged. What is the probability that Tom will receive a token?

Solution.

The probability that Tom receives a token, p(T), *is the probability that Jack's throw is higher. The sample space consists of* $6 \times 6 = 36$ *different pairs of throws, as seen in the figure below. Tom will receive a token whenever the second number in a pair is greater than the first, as shown in the shaded triangle. Thus* $p(T) = 15/36 = 5/12$.

6	(1, 6)	(2, 6)	(3, 6)	(4, 6)	(5, 6)	(6, 6)
5	(1, 5)	(2, 5)	(3, 5)	(4, 5)	(5, 5)	(6, 5)
4	(1, 4)	(2, 4)	(3, 4)	(4, 4)	(5, 4)	(6, 4)
3	(1, 3)	(2, 3)	(3, 3)	(4, 3)	(5, 3)	(6, 3)
2	(1, 2)	(2, 2)	(3, 2)	(4, 2)	(5, 2)	(6, 2)
1	(1, 1)	(2, 1)	(3, 1)	(4, 1)	(5, 1)	(6, 1)
	1	2	3	4	5	6

2.11 - Under the same conditions as those above, what is the probability that:

a) the sum of the throws equals 8;

b) the sum equals 7 or 11?

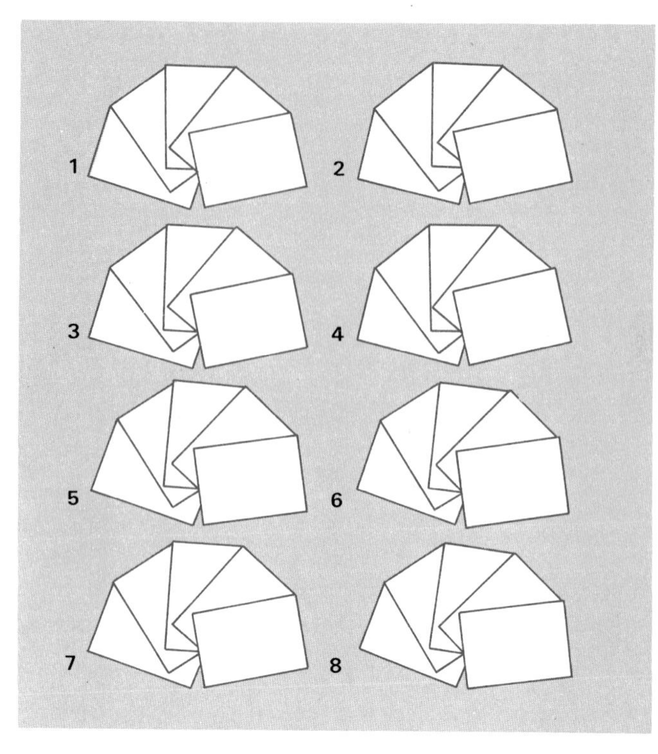

Above: Some combinations in poker: 1, royal flush; 2, four aces; 3, full house; 4, flush; 5, straight; 6, three of a kind; 7, two pairs; 8, one pair. For each combination we can calculate the statistical probability (relative frequency). An experienced player does not simply trust chance, but makes probability assessments.
Below: A table with the combinations and the chances of being dealt them at the beginning of draw poker, which is played with all 52 cards of the pack.

Hand	*Possible Number*	*Probability of Receiving it*
Royal Flush	4	1 in 649,740
Straight Flush	36	1 in 72,193
Four of a Kind	624	1 in 4,165
Full House	3,744	1 in 694
Flush	5,148	1 in 505
Straight	10,200	1 in 255
Three of a Kind	54,912	1 in 47
Two Pairs	123,552	1 in 21
One Pair	1,098,240	1 in 2.5

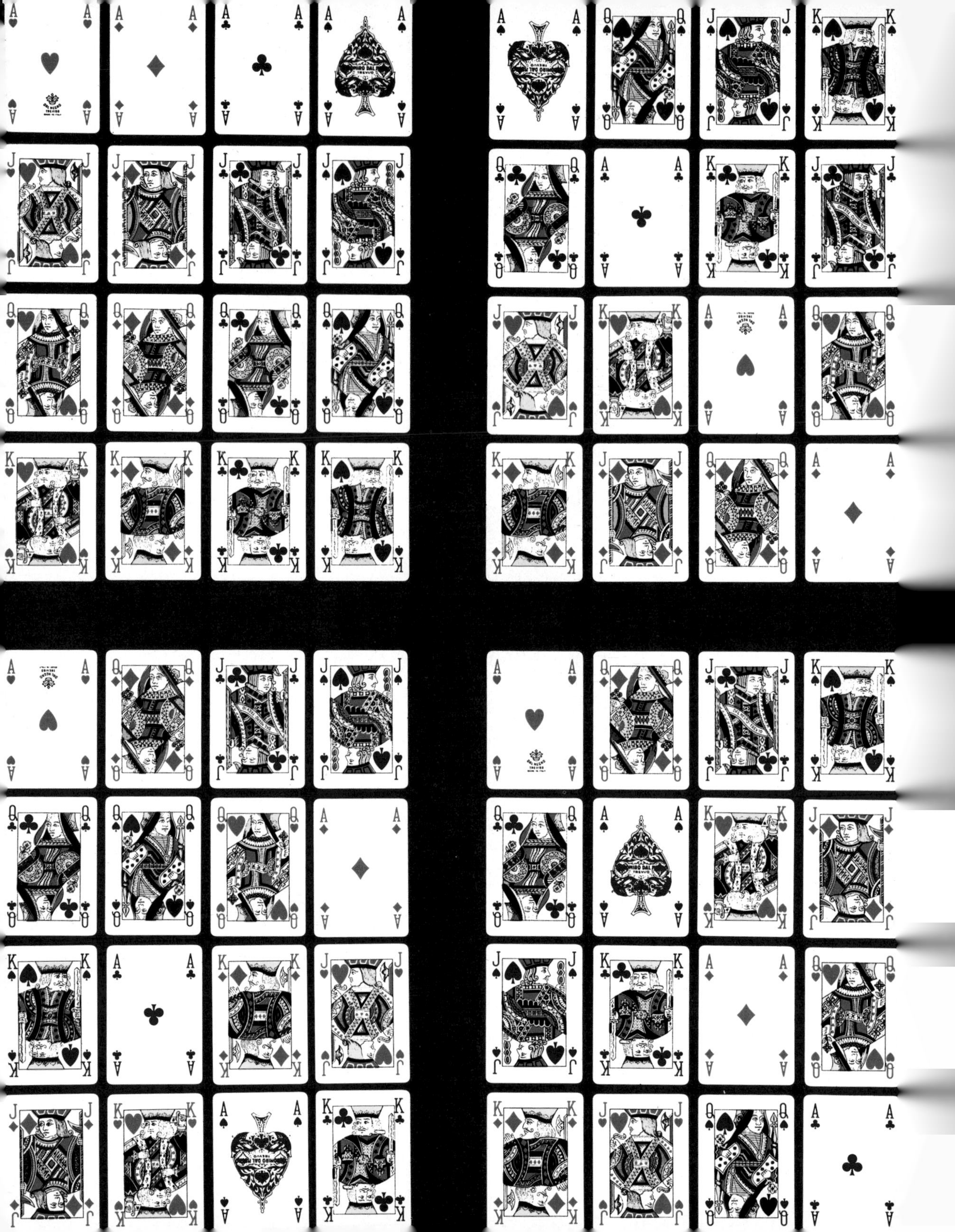

Some card games really belong to Chaper 2 on games with figures. This is one of them: Take aces, kings, queens and jacks (1) and arrange them so one of each appears in each row and column (2), or each colour in each row and column (3), or both (4).

Solution.
Let D = *the sum is 8,* Q = *the sum is 7,* E = *the sum is 11. These events are shown by the three shaded areas in the diagram below.*

Now D = {(2, 6), (3, 5), (4, 4), (5, 3), (6, 2)}, *so that* p(D) = $\frac{5}{36}$. *Similarly,* p(Q) = $\frac{6}{36}$ = $\frac{1}{6}$, *and* p(E) = $\frac{2}{36}$ = $\frac{1}{18}$. *Finally* $p(Q \cup E) = p(Q) + p(E) = \frac{1}{6} + \frac{1}{18} = \frac{4}{18} = \frac{2}{9}$.

6	(1, 6)	(2, 6)	(3, 6)	(4, 6)	(5, 6)	(6, 6)
5	(1, 5)	(2, 5)	(3, 5)	(4, 5)	(5, 5)	(6, 5)
4	(1, 4)	(2, 4)	(3, 4)	(4, 4)	(5, 4)	(6, 4)
3	(1, 3)	(2, 3)	(3, 3)	(4, 3)	(5, 3)	(6, 3)
2	(1, 2)	(2, 2)	(3, 2)	(4, 2)	(5, 2)	(6, 2)
1	(1, 1)	(2, 1)	(3, 1)	(4, 1)	(5, 1)	(6, 1)
	1	2	3	4	5	6

2.12 - Given a bowl with 6 white and 4 red balls, 2 balls are withdrawn consecutively and are not replaced. If the first ball is red, what is the probability that the second ball will also be red?

Solution.
After the first draw, 6 white and 3 red balls remain, therefore the required probability is $3/9 = 1/3$.

2.13 - Two cards are drawn from a standard 52-card deck. What is the probability that both cards will be aces given that:

a) the first card is returned to the deck

b) it is not returned to the deck?

Solution.
The sample space consists of 52 cards. Let A_1=*ace on first draw,* A_2=*ace on second draw.*

a) *If the first card is put back, the sample space is the same for both draws and the two events are independent. Thus* $p(A_1 \cap A_2) = p(A_1)p(A_2) = (4/52)^2 = 1/169$.

b) *If the first card is not put back, the second sample space consists of 51 cards with only 3 aces. Hence* $p(A_1 \cap A_2) = 4/52 \cdot 3/51 = 1/221$. *Note that* $p(A_2/A_1) = 3/51$, *an application of the rule* $p(A_2/A_1) = p(A_1 \cap A_2)/p(A_1)$ *for dependent events.*

2.14 - Three consecutive draws are made from a bowl with 6 red, 4 white and 5 blue balls. What is the probability that the balls drawn will be in the order of red, white, and blue? Take the cases:

a) with replacement,

b) without replacement.

Solution.
The sample space is $S = \{6r, 4w, 5b\}$. *Let* R = *red on first draw,* W = *white on second draw,* B = *blue on third draw. We want to evaluate* $p(R \cap W \cap B)$.

a) *With replacement, the sample space remains constant, so we have* $p(R \cap W \cap B) = p(R)p(W)p(B) = 6/15 \cdot 4/15 \cdot 5/15 = 2/5 \cdot 4/15 \cdot 1/3 = 8/225$.

b) *Without replacement, the sample space changes to* $\{5r, 4w, 5b\}$ *for* W *and to* $\{5r, 3w, 5b\}$ *for* B, *making the probability* $6/15 \cdot 4/14 \cdot 5/13 = 4/91$, *namely,* $p(R \cap W \cap B) = p(R)p(W/R)p(B/(W \cap R))$.

2.15 - A defective die gives equal probability to even throws, each of which is twice as likely as an odd throw. What is the probability that a throw is:

a) even,

b) prime,

c) odd,

d) odd and prime?

Solution.
Let E = {*even throw*}, P = {*prime throw*}, D = {*odd throw*}, Q = {*odd prime throw*}. *We know that* p(E) = 2p(D). *Now* $S = E \cup D = \{1, 2, 3, 4, 5, 6\}$ *and* p(S) = p(E) + p(D) = 3p(D) = 1, *so* p(D) = ⅓ *and* p(E) = ⅔. *Therefore each odd number has a probability* p(D) ÷ 3 = 1/9, *and each even number* p(E) ÷ 3 = 2/9. *Next*, P = {2, 3, 5} *so* p(P) = p(2) + p(3) + p(5) = 4/9 *and* Q = {3, 5} *so* p(Q) = 2/9.

2.16 - In a school 25 percent of the pupils failed mathematics, 15 percent failed chemistry and 10 percent failed both. Select a pupil at random.

a) If he has failed chemistry, what is the probability that he has also failed mathematics?

b) If he has failed mathematics, what is the chance he has failed chemistry too?

c) How likely is it that he has failed at least one?

Solution.
Let M = {*failed in math*}, C = {*failed in chemistry*}, $M \cap C$ = {*failed in both*}. *We know* p(M) = 0.25, p(C) = 0.15, $p(M \cap C) = 0.10$.

a) $p(M/C) = p(M \cap C)/p(C) = 0.1/0.15 = 2/3$.

b) *similarly* $p(C/M) = 0.1/0.25 = 2/5$.

c) *Using the formula on p. 152*, $p(M \cup C) = p(M) + p(C) - p(M \cap C) = 0.30$.

2.17 - A coin is thrown three consecutive times. Consider these events: *A* = heads on the first throw, *B* = heads on the second throw, *C* = tails on the last two throws. Calculate *p*(*A*), *p*(*B*), *p*(*C*) and determine if the events are independent in pairs.

Solution.
The sample space is S = {(hhh), (hht), (hth), (htt), (thh), (tht), (tth), (ttt)}

A = {(hhh), (hht), (hth), (htt)} *and* p(A) = 4/8 = ½

B = {(hhh), (hht), (thh), (tht)} *and* p(B) = 4/8 = ½

C = {(htt), (ttt)} *and* p(C) = 2/8 = ¼

Now $A \cap B$ = {(hhh), (hht)} *and* $p(A \cap B) = 2/8 = 1/4 = p(A)p(B)$.
Similarly, $A \cap C$ = {(htt)} *and* $p(A \cap C) = 1/8 = p(A)p(C)$.
Finally, $B \cap C = \varnothing$ *and* $p(B \cap C) = 0 \neq p(B)p(C)$.
Hence A, B *and* A, C are independent events, but not B, C.

2.18 - Consider the throw of a true die, and these events: *A* = {1, 2, 3, 4}, *B* = {4, 5, 6}, *C* = {2, 4, 6}. Are these independent in pairs?

Solution.
$A \cap B$ = {4} *and* $p(A \cap B) = 1/6$. *Now* p(A) = 4/6 = ⅔, p(B) = 3/6 = ½ *so* p(A)p(B) = ⅔ · ½ = ⅓ *and* A,

B are not independent. $A \cap C = \{2,4\}$ *and* $p(A \cap C) = 2/6 = 1/3$, $p(C) = 1/2$ *so* $p(A)p(C) = 1/3$ *and* A, C *are independent.*

$B \cap C = \{4, 6\}$ *and* $p(B \cap C) = 2/6 = 1/3$, *while* $p(B)p(C) = 1/2 \cdot 1/2 = 1/4$ *and* B, C *are not independent.*

2.19 - Two people, *A* and *B*, practice archery. The probabilities of their hitting the target are $p(A) = 1/4$ and $p(B) = 2/5$. When they shoot at the same time, what is the probability of one of them hitting the target?

Solution.
As the two events are independent, we have $p(A \cap B) = p(A)p(B)$. *The required probability is then* $p(A \cup B) = p(A) + p(B) - p(A \cap B) = p(A) + p(B) - p(A)p(B) = 0.25 + 0.40 - 0.25 \times 0.40 = 0.25 + 0.40 - 0.10 = 0.55 = 11/20$

2.20 - A box contains 8 balls marked 1 to 8. Consider the following draws: $A = \{1, 2, 3, 4\}$, $B = \{2, 4, 6, 8\}$, $C = \{3, 6\}$. Are they mutually independent?

Solution.
$p(A) = 4/8 = 1/2$, $p(B) = 1/2$, $p(C) = 2/8 = 1/4$.
$A \cap B = \{2, 4\}$ *and* $p(A \cap B) = 1/4 = p(A)p(B)$, *so* A, B *are independent.*
$A \cap C = \{3\}$, *and* $p(A \cap C) = 1/8 = p(A)p(C)$ *so* A, C *are independent.*
$B \cap C = \{6\}$ *and* $p(B \cap C) = 1/8 = p(B)p(C)$ *so* B, C *are independent.*
However, the three events are not an independent triplet, for $A \cap B \cap C = \emptyset$, $p(A \cap B \cap C) = 0$, *while* $p(A)p(B)p(C) = 1/16$.

2.21 - In an airplane, the probability of a defect in the automatic landing gear is 10^{-7}. So, too, is the probability of a defect in the fuel supply mechanism. If these two are statistically independent, what is the probability of at least one of them occurring?

Solution.
Call the defects A *and* B. *We need* $p(A \cup B) = p(A) + p(B) - p(A)p(B) = 2 \times 10^{-7} - 10^{-14}$, *which is roughly equal to* 2×10^{-7}, *since* 10^{-14} *can be neglected.*

LIST OF MAIN SYMBOLS USED

U	universal set
$\varnothing$	empty set
A, B, C	propositional variables
$\urcorner$	negation
A'	set corresponding to proposition A
$\overline{A}'$	complementary set to A', corresponding to $\urcorner A$
$\rightarrow$	implication (between propositions)
$\therefore$	"therefore"
$\subset$	proper inclusion (between sets)
$\subseteq$	inclusion (between sets)
$\leftrightarrow$	mutual implication (between propositions), or equivalence
$=$	equality between sets (corresponding to equivalence)
$\wedge$	conjunction of propositions
$\cap$	intersection of sets (corresponding to conjunction of propositions)
$\vee$	disjunction of propositions
$\cup$	union of sets, corresponding to conjunction of propositions
$\dot{\vee}$	exclusive disjunction
$\dot{\cup}$	exclusive union of sets, corresponding to exclusive disjunction between propositions
$/$	incompatibility of propositions
$\in$	"belongs to," of element to set
$\{0, 1\}$	the set of truth-values for categorical propositions
$\leq$	less than or equal to (between numbers)
$\geq$	greater than or equal to (between numbers)
$\neq$	different from (between numbers)
S	sample space set

BIBLIOGRAPHY

Recreational Mathematics

Ball, Walter W., and Coxeter, H. S. *Mathematical Recreations and Essays*. 12th ed. Toronto: University of Toronto Press, 1974.

Gardner, Martin, ed. *Scientific American Book of Mathematical Puzzles and Diversions*. New York: Simon & Schuster, 1963.

———, ed. *Second Scientific American Book of Mathematical Puzzles and Diversions*. New York: Simon & Schuster, 1965.

Kasner, E., and Newman, J. *Mathematics and the Imagination*. New York: Simon & Schuster, 1962.

Steinhaus, Hugo. *Mathematical Snapshots*. 3rd American ed., rev. & enl. New York: Oxford University Press, 1983.

Historical and Philosophical Aspects

Bell, Eric T., *Men of Mathematics*. New York: Simon & Schuster, 1937.

Boyer, Carl B. *History of Mathematics*. New York: John Wiley & Sons, 1968.

Freudenthal, Hans. *Mathematics Observed*. New York: McGraw-Hill, 1967.

Hofstadter, Douglas. *Gödel, Escher, Bach: An Eternal Golden Braid*. New York: Basic Books, 1979.

Kline, Morris. *Mathematics: A Cultural Approach*. Reading, Mass.: Addison-Wesley, 1962.

———, intro. by. *Mathematics: An Introduction to Its Spirit and Use: Readings from Scientific American*. San Francisco: W. H. Freeman, 1979.

Tietze, Heinrich. *Famous Problems of Mathematics*. 2nd ed. New York: Graylock Press, 1965.

Waismann, Friedrich. *Introduction to Mathematical Thinking: The Formation of Concepts in Modern Mathematics*. New York: Ungar, 1951.

Chapter 1. Arithmetic and Algebra

Beiler, Albert H. *Recreations in the Theory of Numbers*. New York: Dover, 1964.

Dantzig, Tobias. *Number: The Language of Science*. 4th ed. New York: Free Press, 1967.

Dilson, Jesse. *The Abacus: A Pocket Computer*. New York: St. Martin's Press, 1969.

Ore, Oystein. *Number Theory and Its History*. New York: McGraw-Hill, 1948.

Chapter 2. Geometry and Topology

Andrews, S., et al. *Magic Squares and Cubes*. 2nd ed., rev. & enl. New York: Dover, 1960.

Arnold, B. H. *Intuitive Concepts in Elementary Topology*. Englewood Cliffs, N.J.: Prentice-Hall, 1963.

Berge, Claude. *The Theory of Graphs and Its Applications*. New York: John Wiley & Sons, 1962.

Crowell, H. R., and Fox, H. R. *Introduction to Knot Theory*. 4th ed. New York: Springer-Verlag, 1977.

Greenberg, Marvin J. *Euclidean and Non-Euclidean Geometries: Development and History*. 2nd ed. San Francisco: W. H. Freeman, 1980.

Matthews, W. H. *Mazes and Labyrinths: Their History and Development*. New York: Dover, 1970.

Nourse, James G. *The Simple Solution to Rubik's Cube*. New York: Bantam Books, 1981.

———. *Simple Solutions to Cube Puzzles*. New York: Bantam Books, 1981.

Saaty, Thomas L., and Kainen, Paul C. *The Four-Color Problem: Assaults and Conquest*. New York: McGraw-Hill, 1977.

Wilson, Robin J. *Introduction to Graph Theory*. Edinburgh: Oliver & Boyd, 1972.

Bibliography

Chapter 3. Set Theory

Halmos, P. R. *Naive Set Theory*. New York: Springer-Verlag, 1974.
Lipschutz, Seymour. *Set Theory and Related Topics*. Schaum's Outline Series. New York: McGraw-Hill, 1964.
Sondheimer, Ernst, and Rogerson, Alan. *Numbers and Infinity: A Historical Account of Mathematical Concepts*. New York: Cambridge University Press, 1981.
Stoll, Robert R. *Set Theory and Logic*. New York: Dover, 1979.

Chapter 4. Logic

Carroll, Lewis. *Symbolic Logic and The Game of Logic*. New York: Dover.
Crossley, J. N., et al. *What Is Mathematical Logic?* New York: Oxford University Press, 1972.
Gardner, Martin. *Logic Machines and Diagrams*. 2nd ed. Chicago: University of Chicago Press, 1982.
Mendelson, E. *Boolean Algebra and Switching Circuits*. Schaum's Outline Series. New York: McGraw-Hill, 1970.
Quine, W. V. *Methods of Logic*. 4th ed. Cambridge, Mass.: Harvard University Press, 1982.
Tarski, Alfred. *Introduction to Logic*. 2nd rev. ed. New York: Oxford University Press, 1954.

Chapter 5. Probability and Statistics

Borel, Emile. *Elements of the Theory of Probability*. Englewood Cliffs, N.J.: Prentice-Hall, 1965.
Fisz, Marek. *Probability Theory and Mathematical Statistics*. 3rd ed. Melbourne, Fla.: Krieger, 1980.
Hoel, Paul G. *Elementary Statistics*. 4th ed. New York: John Wiley & Sons, 1976.
Lipschutz, Seymour. *Probability*. Schaum's Outline Series. New York: McGraw-Hill, 1968.
Mosteller, Frederick. *Fifty Challenging Problems in Probability with Solutions*. Reading, Mass.: Addison-Wesley, 1965.

INDEX